CULTIVATING THE WASTELAND

CULTIVATING THE WASTELAND
can cable put the vision back in tv?

by Kirsten Beck

This publication is a joint project of the
American Council for the Arts
and Volunteer Lawyers for the Arts

American Council for the Arts
570 Seventh Avenue
New York, NY 10018

© Copyright 1983 American Council for the Arts
Additional copies available from:
American Council for the Arts
570 Seventh Avenue
New York, NY 10018

Jacket Design: David Skal
Copy Editor: Bruce Peyton

Typography: ILNY Communications and Media Corporation
Printing: Edwards Brothers, Inc.

Manager of Publishing: Robert Porter

Library of Congress Cataloging in Publication Data

Beck, Kirsten
    Cultivating the wasteland.

    Bibliography: p.
    1. Cable television — United States.   I. Title.
HE8700.72.U6B42   1983   384.55′56′0973   83-12257
ISBN 0-915400-34-0 (pbk.)

This publication was made possible by the generous support of the following contributors:

John and Mary R. Markle Foundation
Andrew W. Mellon Foundation
Knight-Ridder Newspapers, Inc.
Samuel Rubin Foundation

Advisory Committee

ACA acknowledges with gratitude the generous assistance of the following individuals, each of whom reviewed specific chapters of the manuscript and provided valued and thoughtful advice.

Cultivating the Wasteland   **Chloe Aaron**, President, Chloe Aaron and Associates

Beyond the 60-Second Solution   **Robert Alter**, President
Cabletelevision Advertising Bureau

The Arts Transformed   **Kirk Browning**, film and television director

Wiring America   **Kay Koplovitz**, President, USA Network

Copyright Fundamentals
Elements of Deal-Making
The Concepts of Negotiation   **Stuart Rekant**, President
Independent Production Resources

Cable's Access Ability
Unscrambling Cable
Franchising
How to Talk Back to Your
Television System   **James D. Rosenberger**, President
Performance Resources

A Second Golden Age?
Culture by Satellite   **Charlotte Schiff-Jones**, President, Schiff-Jones Limited

A Second Golden Age?
Culture by Satellite   **Mary Anne Tighe**, Vice President for Program Development, ABC Video Enterprises

for
Gerry Goodstein
and
Marian and Tage Beck

Contents

Acknowledgements

Many people representing different facets of the cable television field gave generous amounts of their time while I was researching this book. Representatives of the cable services were unfailingly helpful, particularly those working with ARTS, Bravo, and CBS Cable. My thanks, as well, to the members of this book's advisory group: Chloe Aaron, Robert Alter, Kirk Browning, Kay Koplovitz, Stuart Rekant, James D. Rosenberger, Charlotte Schiff-Jones, and Mary Ann Tighe.

There is no way to adequately express my thanks to the American Council for the Arts for standing behind me through the process of writing this book. Bob Porter, in particular, deserves a medal for his patience, his editorial skill, and his fortitude, as together we tackled the world of cable television and the arts.

People in cities where the arts community participated in franchising shared their experience and helped me to convey some of the hope and frustration of that process. As I began to understand the real potential of local cable for the arts, the counsel of Chuck Sherwood, George Stoney, Craig Watson, Brian Owens, and Dan Leahy was exceedingly helpful.

To Merrill Brockway, I express my gratitude for his generosity of spirit. He taught me much about art on television.

Pat Carroll, Mary Ellyn Devery, Don Schoenbaum, James B. McKenzie, Kenneth Rinker, Alison Becker Chase, Gerard Schwarz, and James Barbagallo spent considerable amounts of time sharing their experiences in taking performances from the stage and putting them on television. They all added to my understanding of just how difficult that process is.

Many former CBS Cable executives and others who had been connected in various ways with the service gave hours and hours of interview time for my chapter on CBS Cable. Each person presented a particular point of view; no one gave me the entire picture. Ultimately, the responsibility for the interpretation is mine and no one else's.

In May 1981, New York University's School of the Arts sponsored a conference on cable and the arts. J. Michael Miller, associate dean, produced the conference and approached ACA about publishing and distributing the conference proceedings which I had edited. This book grew from that initial phone call. My gratitude to Michael for taking the first steps and introducing me to Bob Porter. Thanks are due as well to Michael and New York University for granting the use of some portions of the conference proceedings.

The editorial assistance of Charline Allen, a colleague, friend, and neighbor was essential to the first two chapters of the book. In addition, she was always quick with a helpful suggestion when I became stymied by something.

Thanks to David J. Skal for his smashing cover design and for our many provocative discussions about television and art.

Many at ACA were helpful. My thanks to Bruce Peyton for his improvement of my prose. To Joe Ligammari goes credit for the wonderful title of this book, and for many of the chapter titles. Without the dogged photo research of David Kuhn, this book would not be so beautifully illustrated. My thanks to Bill Keens, Mercia Weyand, William Linden, and the entire staff of ACA for their good humored company and for cheering me on when the going got difficult.

My thanks also to Arlene Shuler, executive director of Volunteer Lawyers for the Arts, for raising the funds that helped to make this project possible and for shepherding those who gave their special legal and negotiating expertise to this publication. The contributions of R. Bruce Rich, Richard J. Lorber, and Timothy J. DeBaets provide essential information for people involved in television production.

The financial assistance of the John and Mary R. Markle Foundation, Andrew W. Mellon Foundation, Knight-Ridder Newspapers Inc., and Samuel Rubin Foundation was essential to this book. My gratitude goes to them all.

Katherine Epler, Peggy Skoulis, and Karen Vrotsos each volunteered to give generously of their time to this project, and I am grateful for their help.

Linda Brodsky was the first in a series of people who taught me about the cable industry. Linda guided me through my first experience with cable language and technology. She directed me into Women in Cable, a professional organization, which was a wonderful source of both information and support throughout this project. I was never without a resource, thanks to my fellow Women in Cable members.

Associates who contributed in a special way include: Jean Clarkson, Cecile Kramer, Erna O'Shea, Eileen Pally, Lorye Watson, Meredith Warshaw, Sandy Warshaw, Debbie Weiner, and Sally Sternbach.

This book could not have been written without the special help and support of Gerry Goodstein, friend, editor, reader, and diligent supplier of humor. Together we laughed through many difficult moments. Thanks, too, go to family and friends who continue to care about me despite the fact that I have been absent and preoccupied for the last six months.

Nancy Hanks died while I was finishing this book. I wrote my first book while working for Nancy, and her influence on me continues to this day. I am grateful for what she taught me and very sorry we cannot share the birth of this book.

K.B.

Introduction

I knew that cable television had arrived as a significant player on the cultural scene when *Buck*, a play by Ronald Ribman, opened off-Broadway in March 1983. The play, set in a television studio, excoriates cable, but the fact that the play was written is more important as a cultural signal than the playwright's attitude toward cable. The writing and production of the play signifies that cable has passed rapidly from a topic regularly covered in the business pages to a phenomenon important enough to become the focus for a major contemporary playwright.

Television is the dominant communications medium of contemporary life. Now, with cable in nearly 40 percent of U.S. households, cable too, has become a serious element in the communications picture. Serious in part because cable has pretentions. Some in the industry contend that cable will enrich the "vast wasteland," and that the arts will help.

In 1961, Newton N. Minow, then newly appointed chairman of the Federal Communications Commission, challenged broadcasters "to sit down in front of your television set when your station goes on the air and stay there without a book, magazine, newspaper, profit and loss sheet, or rating book to distract you — keep your eyes glued to the set until the station signs off. I can assure you that you will observe a vast wasteland.

"You will see a procession of game shows, violence, audience participation shows, formula comedies about totally unbelievable families, blood and thunder, mayhem, violence, sadism, murder, western bad men, western good men, private eyes, gangsters, more violence, and cartoons. And, endlessly, commercials — many screaming, cajoling, and offending. And most of all, boredom. . . . And if you think I exaggerate, try it."

Minow was right, and many of us who do not willingly endure boredom stopped watching television. I was one. Then, in 1977, when a television set was brought into my home in New York City, cable was necessary for reception, and subscribing to cable in Manhattan gradually changed my negative attitude. Even then, the array of channels was impressive when compared to the broadcast stations available in my hometown, Washington, D.C. My interest in cable grew over the next few years, and in 1980, cable mania hit New York City. Everybody in the media and arts, it seemed, was "getting into cable." People were leaving jobs they had held for years to work in cable. Others were developing deals with producers. Still others were becoming producers themselves.

That was also the year Ronald Reagan was elected president, and in December 1980, shortly before the new administration proposed a 50 percent cut in the budget of the National Endowment for the Arts, Bravo, the first national cultural cable service, was launched. Announcements were soon made of others to follow. Thus, the new cultural services arrived at a time when the arts were sinking on the nation's list of priorities and when arts organizations were wondering how they would replace lost government funds.

Conservatives were whispering that the arts shouldn't worry, cable would step in. In June 1981, the Manhattan Institute, a conservative think tank, invited Lewis Lloyd to present a paper exploring whether the new video companies would provide significant income for the arts. After Lloyd told the truth as he saw it — that cable wouldn't save the arts — he was omitted from the published records of the proceedings. When asked to explain this, the president of the Institute admitted, "We did a disservice to Lloyd by pretending he wasn't even in attendance."

At the same time, conferences sprang up all around the country, articles were written, even whole magazines were devoted to the topic. Special reports, some of them selling for thousands of dollars, were prepared. The question of the day was: Will cable save the arts?

The plan for this book was made in that atmosphere. I was quite sure that cable wasn't going to save the arts, and I had no intention of standing idly by and watching the Reagan conservatives get away with that argument.

During the research and writing of the book, I discovered exactly what I expected to find — that cable will *not* save the arts. But I was genuinely unprepared for the other discovery that I made about cable. While it won't save the arts, cable holds considerable promise in ways we had little reason to suspect in 1981, when Bravo, ARTS, and CBS Cable dominated the thinking about cable television and the arts.

When I began interviewing people outside of New York City, I began hearing about arts-cable activities unrelated to the national services. Artists and organizations were working with local cable companies. I learned about access and local origination channels. In short, I began to understand what local cable television was. And the more I learned, the more excited I became.

In the past, all but the most well-known artists and institutions have been barred from television, because PBS and the networks televise only "names," and the expense of buying commercial time puts television beyond the reach of most others. What I discovered in local cable was a television voice for the arts—one which could be available to nearly anyone.

If there is genuine cause for excitement about cable and the arts, it is because there are new possibilities today. In many places the arts now have access to the most potent communications tool in our society today. This book, which started out to be about national cultural cable services has—ironically—turned out to be about community prospects for cable and the arts.

I don't, however, mean to give the impression that the cultural networks are not important. They are, and they have given me great pleasure as a viewer. They have also shown me aspects of the arts I could not have discovered in any other way. I express my gratitude to the programmers and to the executives who labor (and labored) to help those services survive. They have enriched my life.

I hope this book will help you understand the basic technology of cable television—just how the signal gets from here to there. The book also covers the business of cable. I have tried to elucidate the major considerations of people and companies doing business in this exceedingly complicated industry. I have tried to examine some of the issues involved in programming and, in particular, some of the difficulties and joys of trying to put the arts on television. Finally, I have tried to explore the promise and pitfalls of local cable involvement for the arts. For those organizations that are actually creating cable deals, or anticipate doing so, there is a section written by lawyers on how to proceed. Armed with this information, I hope you will feel prepared to take part in this promising medium.

Kirsten Beck

New York City
June 1983

Wiring America

Barely a day goes by without an article on the cable television industry or cable technology in one of the country's major dailies. Since 1979, nearly every sizable communications corporation in America—including ABC, Westinghouse, Twentieth Century-Fox, The New York Times Company, and Reader's Digest—has announced its entry into the cable television field. Why so much fuss over a technology that has been around since 1948?

The so-called video revolution has created nearly as much excitement on Wall Street as the invention of the computer. The wired nation is apparently upon us, offering a bounty of services, such as banking and shopping at home, and entertainment programs as diverse as people's tastes, all available twenty-four hours a day on the more than one hundred channels of different programming which cable is capable of delivering to our television sets.

Underlying all the publicity is the undeniable fact that there is currently more cultural programming available via cable than ever before in the history of television. Does this mean that cable executives have finally realized what many in the arts have known all along—that there really is a market for intelligent and provocative programming to counter the "Laverne and Shirley" mentality of the networks? Though it is too early for a conclusive answer, a review of cable television's origins and its development may provide some insight into where we are today and where we are headed.

The excitement currently surrounding the cable industry could hardly have been envisioned by cable television's inventors when, in 1948, cable was introduced nearly simultaneously in Oregon, Pennsylvania, and Arkansas to spur the sale of television sets. In its earliest phase, cable television was simply an elaborate antenna system designed to carry television signals to homes in areas where geography blocked reception.

Broadcast television signals travel through the air, along lines of sight, and are blocked when they hit mountains or tall buildings. For the same reason, television signals pass directly over the roofs of homes located in deep valleys. Thus, in the early days of television, set sales on the "wrong" sides of moun-

Workmen stringing cable on utility poles. (Photo: Courtesy of Storer Communications.)

tains and in valleys were understandably limited. Consequently, some of cable television's pioneers were appliance store owners anxious to create a new market for television sets.

The original cable television operations, known as Community Antenna Television, or CATV, consisted of tall antennas erected on high ground to "catch" broadcast signals and transmit them by wire cable to subscribers' homes. The cable was strung from trees, utility poles, lampposts, or whatever else was available to carry the signals from home to home, where the hook-up was performed for an installation charge, and the subscriber paid a monthly fee thereafter for the CATV service.

Thus, cable systems brought the new medium of television to people who had previously been unable to receive it, and created a market for television sets where none had existed. Though the beginnings of the cable industry may now seem inauspicious in contrast to its current boom, a story often told in cable circles illustrates what a momentus event it was to receive television for the first time: The town of Lansford, Pennsylvania declared a school holiday the day the cable system was turned on in 1949.

The FCC Steps In

The activities of the Federal Communications Commission (FCC) from 1948 to 1952 contributed greatly to the growth of CATV services. Established by the Communications Act of 1934, the FCC was assigned the responsibility for licensing radio and television broadcasters. Because of the limited availability of spectrum space usable for broadcast, the government viewed the use of the space as a privilege to be granted in the public interest and retained the authority to allocate the space. The FCC's jurisdiction, therefore, included broadcasting but not cable, since it was not a broadcast medium.

In 1948, swamped by applications for licenses to operate new television stations, the FCC froze licensing. This freeze limited the number of broadcast stations to approximately one hundred and remained in effect until 1952.

Because television was new and a source of home entertainment, there was enormous demand for it. Everybody wanted to see "I Love Lucy" and "Howdy Doody," but people not served by broadcasters licensed before the freeze were out of luck. For these people, cable television provided an alternative. Wherever an antenna could be constructed to pick up a signal and a cable strung, television could reach these new markets. In effect, the FCC's action encouraged cable's growth.

The advent of microwave transmission (point-to-point, line-of-sight transmission at extremely high frequencies) in the late fifties and early sixties brought the next major change in cable development. Microwave transmission travels much farther than ordinary broadcast transmission, making it possible for cable systems to provide broadcast signals from distant towns to isolated areas. With this "importation of distant signals," as it was called, the available programming became more diverse. Clear reception was no longer the only reason to subscribe to cable television.

As cable service became more attractive, broadcasters began to realize that it presented a potential competitive threat. That perceived threat was the basis for a long battle waged in the courts, in Congress, and before the FCC to slow down or prevent further growth of cable television in markets that were already available to over-the-air broadcasters.

The FCC, apparently anxious to protect the financial health of the then-young broadcast industry (particularly that of the UHF stations, many of which claimed they were in perilous financial straights), stepped into the picture and issued regulations covering cable systems using microwave to import distant signals. Although the cable operators fought the FCC in court, in 1965 the Supreme Court upheld the FCC's authority, and in 1966 the FCC assumed regulatory jurisdiction over all cable systems. Any operator in a top-100 television market had to apply to the FCC for permission to carry a distant signal. A virtual avalanche of applications resulted, and in 1968, the FCC froze the cable operator's right to import distant signals into major markets. Cable's urban growth was effectively stopped, and by the time the freeze was lifted in 1972, the cable industry's momentum in all areas had been slowed.

With the lifting of the freeze, the importation of distant signals again was permitted within certain limits, marking the beginning of a general trend away from regulation.

The Satellite Era

At the beginning of the seventies, cable television was still primarily a service providing clear television reception and some additional programming. The revolution (and profits) that many had envisioned had failed to materialize, and the industry was jolted by several near-bankruptcies.

The cable revolution—if there really is one—began with the addition of one key element to the technological mix of the first twenty-five years of cable—the satellite. Up to this point, cable operators had made their profits by attracting large

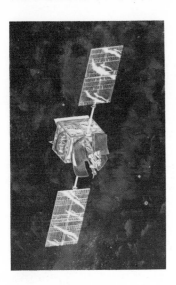

Artist's rendering of RCA's Satcom III-R satellite, which carries twenty-four channels for the cable industry. (Illustration: Courtesy of RCA News and Information.)

numbers of subscribers in areas where over-the-air reception was poor and it was not too difficult to string cable. Once these areas had been "wired," how was the operator to expand? The importation of distant signals offered new programming, and this helped to attract additional subscribers. But by the early 1970s, the cable market seemed to have reached a plateau.

Cable companies in smaller communities, where the movement started, were still earning steady incomes, but the expansion that many had hoped would bring heady profits was not possible until the cable companies could stimulate urban market interest. Most major cities had comparatively bountiful program choices and adequate reception, so what sort of bonus would attract enough subscribers to make the cable companies' enormous initial investment in stringing cable worthwhile?

Domestic satellites provided the bonus the cable industry needed. In 1973, Teleprompter, a major cable operator, and Scientific-Atlanta, a manufacturer of communications equipment, put on a satellite communications demonstration for a National Cable Television Association (NCTA) meeting in Anaheim, California. This revolutionary event beamed a speech by Speaker of the House of Representatives Carl D. Albert from Washington, D.C. to the NCTA delegates via satellite. "The spacecast was as clear as if it had come from next door, and the cable operators got the same point: Television programs could make the same dizzying leap," explained Ralph Tyler in *On Cable* magazine.

Relaying signals long distances via microwave was difficult and costly. Relaying signals via satellite was less difficult, but still costly. It was Home Box Office (HBO), owned by Time Inc., which demonstrated that satellite distribution of programming was feasible. HBO, a pay television service which had been distributing movies and sports events via microwave, decided to convert to satellite distribution. HBO leased space on Satcom I (a domestic communications satellite capable of receiving and retransmitting signals from earth) and convinced a number of cable operators to acquire the receiving "dishes" (antennae) to receive the signal. This required a new and relatively substantial capital investment on the part of the cable operators, but the additional income generated by subscriber fees paid for the HBO service helped to defray that cost. Here was a way for the cable operator to expand the business and increase the profits.

The joining of these two technologies — cable and satellite — has created a new period in cable's development, again stimulating discussions of a cable-induced revolution in com-

munications. We hear of satellite distribution of up to sixty-five channels of programming. Those who wonder why anyone would consider sixty-five different channels of today's television revolutionary are not alone. However, what *is* revolutionary is the diversity of programming and services that is possible. At this moment, the cable industry is bent on exploring and exploiting the technological capabilities of the elements of the cable system. But looking ahead to the intersection of the satellite, the data bank, the home computer, and the interactive cable system, we can indeed see the prospect of a wired nation.

The Hardware of a Cable System

The physical plant of a cable system consists of dishes, receivers, amplifiers, converters, and descrambling devices. These elements work together in a system to distribute a television signal via coaxial cable or optical fiber.

The signal that travels over the cable can originate in a number of different ways:

· It may be broadcast over the air by a commercial or PBS television station and then picked up by the cable system's antenna.

· It may be beamed from a satellite and "caught" by the cable system's *dish* (a parabolic antenna, called a "dish" because of its shape).

The complex "life cycle" of cable television is portrayed above. The story begins with the network headquarters at top sending their programs by landline to an uplink, from which they are transmitted to communications satellites. The programming is sent down to earth stations connected to a cable system's head end, or electronic control center. Also coming into the head end are programs delivered by microwave (at left), from local broadcasting stations and directly from the system's studio. The fare then goes out to the subscriber's homes by cable. (Illustration: Reprinted with permission from the January 1982 *ON CABLE Magazine* © 1982.)

The master control room of Warner Amex Cable in Pittsburgh. (Photo: Courtesy of Warner Amex/QUBE.)

• The signal may originate as a microwave signal or be relayed over long distances via microwave to be picked up by the system's microwave receiving dish (as is done with some distant signals mentioned earlier).

• It may originate in the cable system's own studio as locally originated programming, public access programming, or any one of a number of different automated alpha-numeric services (information such as news headlines and financial market data presented in words and numbers only using a format resembling the printed page).

No matter where they come from, all signals are fed into a central point, called the *headend* of the system, where they are cleaned, amplified, assigned frequencies, and funneled into the main coaxial cable (also called the trunk cable) to be carried to subscribers. As the signal travels along the coaxial cable, it is amplified so that it remains strong and clear until it reaches the subscriber's home.

Some cable systems can carry over one hundred channels of programming, many more channels than can be delivered over the air by broadcast signals. Coaxial cable ranges from one-quarter to one inch in diameter and consists of a core of copper surrounded by an insulation material encased in seamless aluminum and copper shielding. Coaxial cable's extremely effective shielding from outside interference creates a controlled transmission medium capable of carrying many signals. This is in contrast to broadcast technology, where the amount of spectrum space available for television is limited and the signals broadcast over the air need more "room" in which to travel.

Once the cable enters a subscriber's home, it is connected to a converter. In some cases, mostly older systems, the cable is attached directly to the television set. The converter, usually a small box that sits on or near the television set, re-processes signals to a frequency the television can accept and serves as the television's tuner. A descrambling device may also be part of the converter box to decode the signal for a pay television service — such as HBO, Bravo, or Showtime — for which subscribers pay an extra monthly fee (in addition to the monthly charge for "basic" cable services).

Satellites: How They Deliver Programs

Satellite distribution of programs is an elegantly simple concept, with a satellite and any number of earth stations as the basic elements of the operation. Programming in the form of an electronic signal is beamed from an earth station (a parabolic antenna also known as an *uplink*) to the satellite, where it is amplified, switched to a different frequency, and re-transmitted to earth. There it can be received by any dish (a parabolic antenna, known as a *downlink, earth station*, or *TVRO*, short for television receive only) that has been adjusted to receive those signals.

Because each satellite requires a differently aligned dish, a cable system must have as many dishes as there are satellites from which programming is being received. Each dish "collects" the signals sent to it from a particular satellite; then the signals are fed into the system's headend receiver, processed, and sent out by cable to subscribers.

Satellites are placed in a fixed orbital position (geosynchronous orbit) and transmit signals to a defined area of the earth, referred to as their *footprint*. Satellites used for cable communications must be located over the equator between approximately 70 degrees and 143 degrees west (see illustration). Placement of a satellite outside these parameters would mean the footprint would not reach the desired area.

The space available on a satellite for signal transmission is limited, since each signal requires a separate transponder (the hardware that actually receives the signal, amplifies it, changes its frequency, and retransmits it to earth). The satellites currently in operation each have only twenty-four transponders. Because of the growing popularity of satellite

A parabolic antenna, commonly referred to as a *dish*, receives the signal beamed from a satellite. (Photo: Courtesy of Burnup & Sims.)

SATCOM
5
143° W

SATCOM
1
135° W

SATCOM
3R
131° W

COMSTAR
4
127° W

WESTAR
5
123° W

SATCOM
2
119° W

ANIK
3
114° W

ANIK
B
109° W

ANIK
D
104.5° W

WESTAR
4
99° W

COMSTAR
1 & 2
95° W

WESTAR
3
91° W

COMSTAR
3
87° W

SATCOM
4
83° W

WESTAR
1 & 2
79° W

Diagram of satellites presently in use and their positions as they orbit the earth. (Illustration: Courtesy of Westat.)

communications and satellite distribution of programming, the space available for use by the cable industry may become crowded in the future.

More than any other single development in the cable industry, satellite distribution of programming changed the business. Programming had previously been received over the air, relayed by microwave, or taken from tape cassettes plugged directly into the headend. Now, for the first time, all cable systems were potentially linked. One transmission via satellite could reach any cable system in the country with a dish adjusted to receive that signal. Compared to other methods, satellite distribution is extraordinarily efficient, and this relative ease of operation revitalized the industry and reawakened both press and investor interest.

Channel Capacity

In his book *The Wired Nation*, Ralph Lee Smith explained, "Television is a colossal hog of the electronic frequencies. The elbowroom required by each one channel is what makes over-the-air very high frequency (VHF) television spectrum the scarcest of our natural resources. . . . One of cable television's great potentials is its inherent ability to end this . . . scarcity. . . ." Indeed, the possibility of dramatically increasing channel capacity is one of cable's great promises. For the most part, however, it remains *only* a promise.

According to the NCTA, about 45 to 50 percent of all cable systems have a capacity of twelve channels or less, and about 11 percent have thirty channels or more. Both independent cable operators and large companies owning many cable systems operate these twelve or fewer channel systems. Many of them were built in cable's early days, when even as many as twelve channels seemed quite remarkable. By contrast, one of the largest cable companies, Group W Cable, owns a system in Irving, Texas (close to Dallas) with a 107-channel capability, not all of which is being programmed currently. Still, despite all the publicity generated by large corporations battling for franchises in major urban markets and promising fifty to more than one hundred channels, the actual channel capacity for most of the country is nowhere nearly that abundant.

Interactive Cable

An emerging, state-of-the-art development in cable technology is the interactive system, in which signals can travel from the headend to the subscriber (called *downstream*) as well as from the subscriber's converter to the headend (called *upstream*). Interactive cable systems such as Warner Amex's Qube and Cox Cable's Indax can scan the entire system by computer once every several seconds (eight in the Qube system). The computer registers which sets are turned on and to what channels they are tuned.

Interactive systems enable the subscriber to send messages to the headend computer. Subscriber homes in the Qube system, for example, are equipped with a small keypad with buttons which can be pushed to send such "messages." These can be used to respond to poll questions flashed on the screen and to make purchases via shop-at-home services. Warner Amex's Qube subscribers' homes can also be equipped with burglar, smoke and heat-sensing alarm systems as well as medical alert systems, all of which can use the two-way cable connection.

The advent of the interactive system is important for several reasons. In many ways, it is a potential labor-saving device for a cable system. From the headend, it can handle the installation and deletion of subscribers' services, such as premium channels, and in some cases troubleshoot for needed repairs. The interactive system also makes it practical and easy for the cable system to sell individual programs on a pay-per-view basis. The computer serves as a remote box office by recording which subscribers are tuned to a certain channel offering a pay-per-view event.

In the future, the two-way communication capability of in-

Children using **QUBE**, the in-
teractive cable system of Warner
Amex. (Photo: Courtesy of Warner
Amex/QUBE.)

teractive systems will also make possible a radical change in
the nature of the services a cable company can offer
subscribers. The promise of banking at home, access to data
banks, electronic security, and certain kinds of home conser-
vation measures are only a few of the services that may be
around the corner. A word of caution: The central computer
which scans the system will be learning a great deal about
subscribers in the fully interactive system. Some states and
municipalities are addressing this question now by drafting
cable privacy legislation.

The Business of Cable

Discussions of the cable business are often confusing to
those unfamiliar with the cable language. In concept, if not in
practice, the cable industry can be separated into three functional parts: the cable system operator, the program supplier,
and the equipment supplier. A second way to distinguish the
functions in the cable business is to differentiate between the
hardware and software sides. Hardware comprises the cable
plant and equipment, installation, and operation. Software
comprises programming and programming suppliers.

An understanding of the cable business begins with *the cable
system operator*, the company that holds a franchise to provide
cable television to the residents of a certain locality. The cable
system operator is responsible for the building, operation,
and maintenance of the system: the headend, the cable and its
components, and the receiving equipment in subscribers'
homes.

The cable system operator (often called *the operator*) may be
a small company like the ones that originated CATV services.
Or the operator may be part of a larger company, known as a
multiple systems operator (MSO). An MSO is a company that

owns more than one cable operation. Often these are located in different states. Because of the investment appeal of the cable business, major corporations such as Time Inc., American Express, and The New York Times Company have moved into cable system operation, acquiring many smaller systems. This sort of acquisition created the MSO and changed the original character of the industry.

In many cases, the cable operator originates either little or no programming, instead drawing it from a variety of sources. The FCC requires the operator to carry certain "local" over-the-air broadcast signals. Some operators import distant signals as well: WTBS, Ted Turner's Atlanta superstation, for example, or WOR from New York City. Operators sometimes run alpha-numeric services, which give weather reports, sports results, news headlines, stock market information, and other automated information.

Locally originated programming may come from a number of sources. Access is time given without charge by the operator (usually on a first come, first served basis) to individuals, community groups, local governments, and service institutions. Some systems also lease time for programs which individual producers wish to create and cablecast. The producer of leased access programming can sell ads to defray the cost of creating the program. Some systems create their own (local origination) programming and sell ads, using this as an additional revenue stream.

Because local governments control access to the public rights of way which cable operators use for laying or stringing cable, a municipality has the power to choose which cable company is given the right to service the community. This is called granting a *franchise*. In most cases, as a condition of the franchise, the cable operator is required to pay the municipality a fee — normally a set percentage of gross revenue.

The increasingly heated competition for franchises has also enabled municipalities to seek services in addition to franchise fees. These may include *access* channels — reserving certain channels for public use — with the cable company providing studios, equipment, and production staff. Some franchise agreements also set aside channels especially for governmental, educational, or arts uses, for example. Arts organizations and other community groups have been very active in such franchise negotiations in some locations.

Program suppliers buy, create, or originate programming and sell it to the cable system operators for distribution to subscribers. Suppliers range from Ted Turner, with WTBS, Cable News Network (CNN) and CNN Headline News, to HBO and the Cable Health Network. Such companies are also responsible for getting their program signals to the system operators, which they accomplish by maintaining their own uplinks to transponders leased or bought from satellite common carriers like RCA and Western Union. (Transponders are maintained by the satellite owners.)

Financial arrangements between program suppliers and cable operators vary. *Basic services*, those for which subscribers do not pay an extra fee, carry advertising which (theoretically) covers the supplier's costs. CBS Cable and Hearst/ABC's ARTS, for example, were both planned as services that would ultimately be supported by advertising revenues. However, as of this writing, none of the basic "ad-supported" services is in the black, and CBS Cable has gone out of business because of its losses.

Contracts with cable operators for basic services differ, but terms sometimes include a per-subscriber charge to the operator. For example, in early 1983, the USA Network was charging eleven cents per subscriber per month, Nickelodeon fifteen cents, and CNN between fifteen cents and twenty cents. In most cases, the program supplier sets aside a certain amount of time for "local avails" (advertising time that can be sold to local advertisers by the cable operator). In 1982, some basic service suppliers began compensating cable operators for channel space as an inducement for the operators to carry their programming. But this practice is unlikely to persist during the economic squeeze brought on by recession in the eighties.

Pay services (also called *premium services*) such as HBO, Showtime, and Bravo, a cultural pay service, operate differently. For each of these, the subscriber pays an extra fee, a percentage of which is retained by the local cable operator who passes along the balance to the program supplier. The details of these arrangements vary according to the supplier and the individual system, but generally, the retail price paid by the subscriber is roughly double the wholesale price the operator pays.

The High Stakes of Building a Cable System

Two principle factors determine a cable operator's estimated potential revenue from a franchise: first, the number of homes that exist along the cable's route (referred to as the density, or the number of homes per mile of cable) and second, the percentage of homes on the route that actually subscribe to the service (referred to as the penetration).

Despite the fact that the hardware side of the industry is capital intensive, requiring large initial outlays during construction, the original cable systems experienced a reasonably quick return on their investments, prompting some observers to call them cash machines. Once an initial investment had been recouped, a system had only maintenance expenses as it continued to bring in regular monthly subscriber fees. That picture has since changed. In 1979, one MSO estimated the building cost of the average system (including plant, electronics, and labor) at $7,000 per mile. By 1981, that estimate had more than doubled to $15,000 per mile to construct a state-of-the-art system. The same company's 1983 estimates have increased to nearly $20,000 per mile.

Furthermore, in the battle to win urban franchises, some MSOs have made costly promises of equipment, of publicly-available, staffed studios, and of interactive systems with over one hundred channels. Assuming that the companies actually deliver what they have promised, some speak of waiting as long as fifteen years after the initial start-up before showing a profit.

Because this cable game has become so expensive and companies must be able to wait for a return on their investment, the industry is now dominated by major corporations. A small, under-capitalized company cannot sustain the costs of a new franchise. In some cases, as older franchises come up for renewal and the licensing municipalities require the cable systems to be rebuilt and brought up to date, small cable operators are selling out to larger companies rather than face the expense of rebuilding. All of this tends to increase the power of the large MSOs as an industry force.

Program Services: High Costs and Fierce Competition

Companies wanting to start new program services (the software side of the industry), whether subscriber- or advertiser-supported, face their own problems. First among them is the start-up expense, with estimates ranging from $10 million to nearly $100 million. Industry sources report that Walt Disney Productions projected that its new channel would cost roughly $150 million over four years even though it already had a large store of existing programs. These costs contrast sharply with the estimated $500,000 or less that it took to start Showtime in 1976. It took both HBO and Showtime five years to move into the black, however, and as of mid-1983, no other pay cable service is showing a profit.

The cost of creating a new service is not the only problem facing the new programming kid on the block. These days, that new kid is having a hard time finding space to play.

Twelve-channel systems have more than enough programming to fill their needs, and the same holds true for most twenty-four-channel and some thirty-six-channel systems. Indeed, a number of cable operators are turning away program services, and the competition for channel space has become fierce.

In some cases, the major companies that own MSOs are also suppliers of programing. For example, Time Inc. owns one of the largest MSOs (American Television and Communications Corporation, known as ATC) and the most financially successful program supplier, HBO. Warner Amex (a corporate marriage between Warner Communications and American Express) owns an MSO, Warner Amex Cable Communications, and a program supplier, Warner Amex Satellite Entertainment Company.

Finally, as both the consolidation of ownership and the difficulty of securing channel space increase, program services not allied with an MSO may find it more difficult to break into the market. Some observers feel that in the near future it will be financially impossible to start a new service without the assurance of a certain number of subscribers, either to buy the service or to comprise an audience that advertisers want to reach.

The interaction of cable system operators and program suppliers raises questions about the survival of programmers not allied with MSOs. Will this integration close off potential outlets for independent program suppliers? Will the MSOs or their parent companies shut out independent programmers who might compete?

Early in 1982, Cox Cable, with an equity interest in a movie service called Spotlight, dropped Showtime, a competing service, from nine of its systems in order to carry its own offering. Why, after all, should Cox carry Showtime if it could carry Spotlight and earn back a portion of its investment? With channel capacity tight and corporate decisions dictated by clearly defined financial interests in certain program services, more programming decisions will be dictated by corporate ownership. This may ultimately result in closing off some outlets for programming, thereby lessening the choices available to subscribers and increasing a programmer's risks.

A Second Golden Age?

Cable television is exciting to viewers because it promises more diversity than conventional broadcast television offers. Cable not only promises, it delivers.

The Cable Satellite Public Affairs Network (C-SPAN) offers twenty-four hours of public affairs programming and a chance to watch the daily proceedings of the House of Representatives. Not everyone wants to watch such programs, but those who do care passionately and attach great value to a channel that brings the House lawmakers' proceedings directly into their homes.

By the same token, many parents, sensitive to what their children see on television, consider quality children's programs of primary importance. They do not want their children watching violence-laden programs surrounded by ads for sugary foods low in nutrition. First-class children's programming, a hallmark of Nickelodeon, has great value for these discerning parents.

The Playboy Channel, named after the magazine, has attracted equally ardent response. Much of its programming is so-called "soft core," featuring plenty of nudity and R-rated movies. Despite mediocre to bad reviews, viewers in cable system after cable system vote for this pay channel with their pocketbooks.

Given this kind of variety, is it any wonder that the traditional network audience, that mythic monolith, has splintered?

ABC, CBS, and NBC held 92.1 percent of the television audience in the fourth quarter of 1979. By the close of the 1980–81 season, that share had plummeted to 81 percent. Where are these viewers going?

People with cable are watching more television, not less. But they are tuning in something other than the three networks. Between 1980 and 1982, the Nielsen Television Index measured a whopping 15 percent decrease in the networks' share of the prime time audience in homes with cable. (*Share* is the percentage of television-equipped households tuned to a particular channel over a particular time.) With more than 30 percent of the television homes in the

United States wired, it appears that the cable alternative is wooing viewers away.[1]

A major reason the networks are losing portions of their audience is that cabled homes have so much more programming from which to choose — twenty-four hours of movies for movie freaks, twenty-four hours of news for news junkies, an entire channel devoted to health concerns. Programs about, or originated by, virtually every major religion are available, and nearly every ethnic group now has its own special programming. Music Television (MTV), the video rock music channel, and The Nashville Network, new in 1983, appeal to specific tastes in music, and there are two channels devoted to cultural programming. Finally, "blue" programming and "adult" movies round out the groaning board of program selection.

Cable television's potential for plentiful channel capacity, combined with satellite technology, makes such diversity possible. When compared to the three to seven channels commonly available in broadcast markets, cable's multitude of channels changes the viewing environment substantially. It also changes the rules of the television programming game.

Programming used to be based on scarcity, with few choices for the viewer and programmers competing for the largest possible share of the audience, and thus, the largest fees from advertisers. But "largest possible audience" shrinks considerably when viewers can choose from twenty to fifty different channels.

The economics of television programming are different now, too. It is no longer just a race to see who can get the biggest audience. Some cable theorists contend that it is economically feasible to aim programming at relatively small segments of the audience, a practice known as *narrowcasting*. The economic assumption which supports narrowcasting is that such viewers will be so attractive to certain advertisers that they will pay a premium to reach them.

Another programming approach used for cable can be described as *convenience-casting*. By offering the same type of programming twenty-four hours a day — all-news, all-rock-video, all-movies, all-weather — viewers can receive *what* they want *when* they want it, not when the network *says* they can have it. Such convenience-cast programming is similar to all-news radio in inspiration. Some audiences for these services are narrower than others, and the smaller the audience, the less appeal the service will have for advertisers. But those

---

1. According to Paul Kagan, a leading analyst of broadcast and cable television, 33 percent of the U.S. homes with television were wired for cable by the end of 1982. By A.C. Nielsen's count, that figure stood at 37.2 percent as of February 28, 1983.

HBO Theater presentation of *Wait until Dark*, with Joshua Bryant and Katherine Ross. (Photo: Courtesy of HBO.)

advertisers who want to reach rock music fans or news junkies can reach them on MTV, Cable News Network (CNN), and Satellite News Channel at any time of day or night.

Some programming is so attractive to viewers that they will pay extra to receive it. Movie services are examples, and this phenomenon has turned Home Box Office (HBO) into Time Inc.'s biggest money-earner. When subscribers pay extra to receive a service, it is called a *pay or premium service*. There are also so-called *basic* or *advertiser-supported* services, as well as hybrids of the two, pay services that carry advertising.

Pay-per-view is yet another type of programming available. Events that the entertainment industry considers "blockbusters," such as world heavyweight boxing championships and major movie premieres, are currently cablecast to relatively limited audiences as pay-per-view events. As the name implies, pay-per-view programs are purchased singly, per event, from the cable company. (Pay-per-view events are also available over the air from some subscription television services.) For pay-per-view to be economically feasible *for cable*, the cable systems need more sophisticated technology (addressability) than most have now. However, because of the enormous financial promise of pay-per-view, many industry observers expect it to be a common method of program distribution before the end of the decade.

Narrowcasting

The promise of narrowcasting is that television finally will deliver material that may appeal to only a small portion of the audience—cultural and foreign language programs, for example.

Some cable theorists compare narrowcasting to magazine publishing, where a large number of special-interest magazines (such as *American Photographer*), each designed to appeal to a limited but devoted readership, coexist with general-interest magazines (such as *Life*).

For those concerned with the arts, a more disturbing cable analogy is the one sometimes drawn to radio, where stations define themselves as all-news, all-talk, top-forty, all-jazz or country music. Here, the stations with the narrowest audiences—jazz and classical—often find it difficult to survive. Indeed, New York City, a community known for its catholicity of tastes and interests, lost its only all-jazz radio station in September 1980 to an all-country-and-western format.

Although narrowcasting is often cited as one of the most unique features of cable, not all of the industry is committed to narrowcasting. Les Brown, the media observer, contends that even at the beginning of the eighties cable boom, cable people were privately declaring narrowcasting to be a myth. "The people who are running the cable business today are in it for the big bucks," he says. "They want to put the least expensive, most commercial programs on cable." In June 1982, Michael Fuchs, then HBO's executive vice president for programming, told the Television Critics Association, "HBO never said narrowcasting. . . . We did not go on the satellite to narrowcast." And that is clearly true. HBO wants as large an audience as it can possibly garner.

Actually, the cable programming emerging in the early eighties is a mix of services: some straight out of the narrowcasting mold; some mimicking the all-news, all-music radio format; and some seeking the broadest audience possible

The Cable Health Network (CHN), is a classical example of narrowcast programming. Dr. Art Ulene, its founder, says that the cable health service grew out of his frustration with broadcasting restrictions. "On the 'Today Show,' health issues got only three minutes, and you can't tell people what they need to know in three minutes." Out of that frustration grew a twenty-four-hour-a-day service which caters to people who want to know more about health, medical, and science developments. A basic service, CHN offers programming on health and science, keeping fit, healthy relationships,

The Cable Health Network's program "Mommy, Daddy and Me," with John Schuck, Susan Bay Schuck and son Aaron. (Photo: Courtesy of Cable Health Network.)

lifestyles, self-help, medical care, growing up, and getting older. This, of course, is the perfect target audience for vitamin manufacturers and other businesses creating products that cater to the nation's current health and well-being preoccupation. As a result, many such companies are buying time on CHN.

Other services, particularly premium movie services such as HBO and Showtime, are more interested in a broad audience. The broader the audience, the more households there are that will pay the extra fee to receive the service. Indeed, HBO is sometimes called a fourth network. But premium services are not the only ones seeking as broad an audience as possible. Ted Turner has proclaimed his intention to make the broad-based WTBS the fourth network. The USA Network tries to provide something for everyone, very much as broadcast networks do. USA has four program areas: Daytime, geared to women at home during the day; Kidstime, before and after school and parts of each weekend day; Sportstime, geared to male sports fans at night; and Nighttime, late night video music, rock-oriented movies, and concert footage. The USA program divisions draw on the traditional network model of dividing the day into parts (day parts) and programming to specific audiences at different times.

The jury is out on whether programming based on the concept of narrowcasting can survive. Two channels designed to appeal to the narrow audience for "quality" programming (CBS Cable, a basic service, and The Entertainment Channel, a premium service) have already failed. Narrowcast basic program services promise a match between a specific audience and advertisers who want to reach that audience: a

Taping the Entertainment Chan-
nel's award winning production of
Sweeney Todd. (Photo: Courtesy of
RKO Videogroup.)

photography show for photo equipment and film suppliers; a
cultural channel for Mercedes, Jaguar, and caviar distributors; Japanese language programming for makers of Japanese products distributed in this country.

Some *premium* services are narrow in their appeal; Bravo, a
cultural and movie channel, and The Playboy Channel are
examples. Bravo was created as a pay service because, in the
words of its first president Marc Lustgarten, "historically ad
revenues have not supported arts programming. There was
no reason to believe they would now." Bravo's creators,
however, felt that some viewers would find their programming so valuable that they would be willing to pay an extra
fee to receive it. This was also the reasoning behind The
Entertainment Channel, which failed to attract enough
subscribers to continue as a pay service.

No one knows whether narrowcasting in either form will
prove economically viable. However, in order to understand
the considerations involved in choosing a format for a new
channel, be it a narrowcast, convenience-cast, or broadcast
model, it is necessary to understand how program services
are financed, distributed and marketed.

Financing the Programming

The first challenge the programming producer must tackle
is how to assemble the substantial initial investment required
to finance a new satellite program service. Some analysts
estimate it cost the RCA Corporation and Rockefeller Center

Inc. $85 million to start The Entertainment Channel, which failed after only eight months of operation. The same analysts say that Walt Disney Productions is investing only $30 to $40 million in start-up costs for its pay channel. The lower costs for Disney are attributable to the fact that the channel will cablecast only sixteen hours a day, not twenty-four hours as The Entertainment Channel did, and to the large stockpile of software Disney brings to the project. Figures given for the start-up costs of CBS Cable range from $30 to $60 million.

Programming officials prefer not to go on record regarding costs of their services, making precise figures difficult to obtain. But it is clear that the costs of starting a new service — basic or pay — are awesome and climbing higher. Additionally, initial investors must make a substantial commitment to carry a service while it establishes an audience and, if it is ad-supported, a base of advertisers. HBO and Showtime are the only two satellite-delivered program services in the black as of mid-1983, and neither reached the break-even point before its fifth birthday.

Pay services generate revenue through per-subscriber fees charged to the cable operators. The amount the pay service receives is usually about half what the subscriber pays to the cable operator, an average of four to six dollars. Pay services often offer volume or performance discounts to induce operators to promote the service and sell more subscriptions.

Many basic services charge the cable operators a monthly per-subscriber fee, which can range from five to fifty cents, but averages about ten cents. As channel capacity became more scarce in recent years, some services stopped charging and began offering their services free to cable operators (and sweetening the deal with financial and/or promotional assistance at launch time) to convince operators to carry the service.

ESPN (an all-sports basic service), for example, charged a per-subscriber fee at the time it was launched, but abandoned it in 1982 and began compensating systems on a per-subscriber basis to carry the service. That change was short-lived, however, as the service was forced to throw out the compensation plan when advertising revenues failed to live up to expectations. ESPN has now returned to charging a per-subscriber fee.

For basic services, advertising is the major source of revenue, but this revenue has not been accumulating at the hoped-for rate. None of the basic services has managed a sustained period in the black as of this time. Both CNN and USA Network moved into the black in 1982, but were unable to maintain that position.

Although CNN and USA have managed to attract strong advertiser support, the reluctance of advertisers to buy time on other services has caused some basic services to scale back estimated revenues. CBS Cable apparently sold between $8 million and $9 million in advertising (the service had projected it would sell $20 million) in its eleven months of existence. In contrast, USA reported $8 million in ad sales from June to September of 1982.

Sales of programs to ancillary markets provide other sources of income for services that produce their own programming. Home video product companies, airlines, and hotel pay-television services all need programming and are potential markets for programming created by cable companies. Foreign sales can also bring in additional revenue.

Charlotte Schiff-Jones, former vice president for marketing at CBS Cable and now president of her own company, Schiff-Jones, Ltd., says that cable operators are "recognizing that to compete with other video technologies, the quality, quantity, and diversity of service is crucially important. Furthermore, for program services to survive, there must be a triangular system of support with the program suppliers, the cable operators, and the advertisers each occupying one point of the triangle. Until circulation reaches its critical mass and advertising becomes an accepted part of the media mix, the cost of quality programming will just be too high."

By mid-1983, two program services, one basic and one pay, had failed because their losses were too great for their investors. Without an improvement in the economy, a change in the attitude of advertisers towards cable television (which does seem to be beginning), or a change in the financing structure of the programming side of the industry, it seems certain that there will be other failures before a lasting cable programming picture emerges.

Methods of Distributing Programming

Cable programming is distributed to systems in three different ways, determined to a certain extent by the audience sought. Satellite transmission is used for sources which are national in scope, such as CNN, ARTS, and HBO. Other program services are regional in scope — Sportschannel and Prism in the East, for example — and they are distributed regionally via microwave transmissions. Sometimes, tape cassettes are shipped via the mail or package service from one system to another. This method of distribution, called *bicycling* in the industry, is currently the only widely used method to circulate locally produced programming to other systems.

The Marketing of Programming

Since the 1975–76 season, when HBO and WTBS became available via satellite, the number of satellite-delivered services existing or planned has risen to approximately fifty. However, despite the impressive variety of entertainment available, almost no household in the country can view all of it. This is because a particular program or program service must pass successfully through several different markets before it can emerge on the home television set.

The marketing system for cable programming closely resembles the wholesale-retail marketing pattern, in which manufactured goods must navigate a chain of middlemen to reach the consumer. For example, start with a particular program, *Stages: Houseman Directs Lear*, a sixty-minute documentary film by Amanda Pope on The Acting Company's rehearsal and performance of *King Lear*. Pope, as the producer, marketed *Stages* (the product) to CBS Cable (the wholesaler). The film then became part of the package of programs which CBS Cable, in turn, marketed to cable operators (the retailers), each of whom decided whether or not to put CBS Cable on one of their channels (stock the shelves). Finally, the cable operators (retailers) who chose to carry CBS Cable marketed both their system (the stores) and CBS Cable (one of many

"Broadway on SHOWTIME" presentation of *fifth of July*, with Richard Thomas and Swoosie Kurtz. (Photo: David C. Batalsky, Courtesy of Showtime.)

channels/products on their shelves) to the area they served.

Program services face many challenges in marketing to cable operators. The two most daunting are limited channel capacity and limited receiving equipment (a dish that addresses one satellite cannot receive programming from others). Most systems in this country have less than thirty-six channels, and many only twelve.

One of the few FCC regulations still remaining for the cable industry is the so-called *must-carry* rule, which requires that all broadcast stations within an approximate thirty-five-mile radius from the system's headend be carried. A few older systems are grandfathered; that is, they were launched before the rules were implemented. Therefore, they are allowed to carry fewer broadcast stations. Grandfathering usually exempts systems from carriage of duplicated network affiliates. In some major television markets, must-carry rules may require systems to carry as many as nine over-the-air stations in New York, and seventeen in Los Angeles, cutting severely into the systems' channel capacities.

In addition, many local franchise agreements require that access channels be made available to the public, and some franchise agreements include the operator's promise to deliver a particular satellite-delivered channel (or channels). In many cases, this leaves few channels for additional programming. In other words, shelf space for new stock may be extremely limited. *Cable Marketing*, an industry trade journal for cable marketing executives, declared in March 1982 that "the biggest challenge facing programmers...is the sudden abundance of available programming and the comparatively limited channel capacity." Under the circumstances, a cable system's program director "must be more of a businessman than a creative producer of programming."

In considering a new service, the cable operator's program director and manager must weigh a number of considerations. What will the service cost? Will there be a per-subscriber charge? (Ted Turner charges twenty cents for CNN when it, alone, is carried and 15 cents when WTBS is carried along with it.)

Will reception require a new earth station? (CBS Cable, for example, was carried on Westar IV instead of Satcom 3R, which carried the most popular services at the time of the CBS Cable launch. This meant that in order to receive CBS Cable, a cable operator had to acquire a new earth station, an investment of $10,000 to $20,000. Because CBS Cable was anxious to be carried by certain cable systems, it made dishes available to some of the major cable operators.)

Will the cable operator be paid to carry it? (Satellite News

Channel pays twenty-five cents per subscriber in launch assistance, and initially paid an additional fifty cents for "charter affiliates.")

Will it provide marketing assistance to promote the cable operator's system as well as the new service?

Will the new service create *lift?* Will the program service promote *retention?* (Lift is industry jargon for an increase in subscribers, which can result from skillful marketing or because a service is so popular that a cable system gains new subscribers eager to receive it. Retention is the ability of a system to hold onto its present subscribers, based on the appeal of the system's offerings.)

With a pay or premium service, major considerations include whether or not it will complement the other pay services already offered and how the service can be marketed. Operators must evaluate whether or not the service will cause *churn* (subscribers ordering and then disconnecting the service, which results in extra hook-up and disconnection expenses for the operator).

Finally, an operator may consider whether the new offering will provide a public service or enhance the system's image. Nickelodeon, for example, is a respected children's service produced by Warner Amex. However, given the limited channel capacity of some systems, public image and good public relations are frequently a low priority.

With the implementation of more sophisticated marketing techniques by cable operators came the advent of *tiering*, which involves packaging a number of services together and pricing each package, or *tier*, incrementally. Some tiering structures require subscription to an incremental tier in order for particular pay services to be available.

For example, in Warner Amex Cable's Dallas system, Tier I, priced at $2.95, offers broadcast stations, access channels, and a program guide. Showtime and GalaVision are available to Tier I subscribers for $7.45 and $5.95 respectively.

Tier II, priced at $7.50, offers CNN, CNN Headline News, SIN (Spanish Television Network), Nickelodeon, WTBS, ESPN (Entertainment and Sports Programming Network), USA Cable Network, SPN (Satellite Program Network), C-SPAN, ACSN (Appalachian Community Service Network—the Learning Channel), Modern Satellite Network, and MTV. Cinemax, The Movie Channel and HBO are available for $7.45 each; Tier II subscribers, who must also be Tier I subscribers, can subscribe to that tier's pay services as well.

Tier III, priced at $9.95, offers local origination, superstations WGN and WOR, plus five pay-per-view channels, with

Nickelodeon's children's program "Pinwheel," with Jake (George James) and his friends. (Photo: Courtesy of WASEC.)

movies priced at $3.00 to $4.00 per-view. Thus, the Tier III subscriber is actually paying $20.40 per month for basic services, plus additional charges for any pay services or pay-per-view events.

This is just one example of a tiering structure. Obviously, the objective in tiering is to persuade subscribers to pay a slightly higher rate for the more desirable services. Subscribers on the Warner Amex tier, for example, must pay at least $10.45 ($2.95 plus $7.50) to have access to cable-only services like CNN and ESPN, and premium services such as HBO.

This particular approach is reminiscent of packaged-goods stores' "loss leader" approach: luring the customer in with specials on such necessities as bread and milk, with the hope that he will spend money on impulse buys. The cable subscriber is lured by the promise of inexpensive cable service at $2.95 per month, then realizes he must spend considerably more to have the services he most likely wants to see.

A simpler and more widespread approach to tiering usually involves packaging together three to four popular basic program services on top of basic service and charging subscribers an extra fee to see them. CNN, ESPN and MTV, for example, are well-liked among many subscribers, and thus are frequently packaged as part of a tier which may cost the subscriber an additional three to four dollars per month.

As operators adopt even more sophisticated packaging techniques, a few are creating what they call *value-added tiers*, again, usually comprised of three to four popular basic services. A typical offer in a newly built system might be a "package" priced at thirty to forty dollars which includes basic service, at least two pay services (with several from which to choose), a remote control device, a program guide, and a value-added tier. The strategy behind such packaging is to discourage disconnects by either making individual prices for elements of the package very high (that is, this package might cost sixty dollars incrementally), or to tell subscribers that elements such as the value-added tier are not available individually — the subscriber will lose them if one of the pay services is dropped, for example.

Once the cable operator has bought the programming, it must be promoted in order to convince customers to take it off the shelves (to watch the show, or in the case of premium services, to pay the extra fee). Sometimes, programming services provide coop funds (up to a specified amount) to reimburse the cable operator for advertising and promoting the program service to subscribers. CNN, for example, will provide thirty cents per subscriber per system (up to $10,000 per year) when furnished with documentation of a cable system's costs incurred in promoting CNN. Satellite News Channel provides ten cents per subscriber per year, and CBS Cable had provided fifteen cents.

Cable operators will not invest as much time and money in marketing and promoting ARTS as they will in selling such premium services as Showtime and HBO, which offer them a share of each subscriber fee collected. As a result, many potential subscribers within a given market may be relatively unaware of the full range of basic cable services available to them. Sometimes even cable subscribers do not fully understand what programming they receive as a result of their cable subscriptions. (It is not a straightforward task to keep track of programming on more than thirty channels.)

Because of the subscriber confusion created by the multiplicity of channels and programs, many cable operators either subsidize the production of a program guide, or produce their own, as a part of their marketing and public relations operations. Without a guide, all but the most determined subscriber is likely to give up trying to figure out what the system has to offer. These guides give needed publicity to basic services, helping the viewer to sort out one news service from another and giving the schedule of performances on ARTS, for example. Guides also serve as marketing tools for pay services by heavily promoting the most appealing programming carried by the premiums each month.

The increasingly discouraging economy in the early eighties prompted the cable industry to pour more and more of its energies into marketing. With widespread marketing of tiers and the use of complicated formulas for launch and coop assistance, program suppliers and cable operators joined hands with marketing experts from America's packaged goods industries, changing forever cable's early identity as a community antenna service.

Beyond the 60-second Solution

Advertising on cable: Some cable systems are counting on it for additional revenue; some viewers consider it an unexpected annoyance; certain programmers know it is the key to their continued existence. Certainly, for anyone who wants to see cultural programming continue and prosper on cable, advertising is important. Ultimately, lack of sufficient advertising support was a major cause of CBS Cable's failure, and advertising revenues are no less critical to Hearst/ABC's ARTS. Such services, conceived as advertiser-supported, cannot survive without attracting sufficient advertising to cover their costs and profit.

Cable Does Not Mean Commercial-free Television

For many non-cable viewers, the term "cable" is synonymous with commercial-free television. This misunderstanding arises because premium movie services, the most visible of cable's offerings, promote themselves as commercial-free, and people often do not understand the difference between cable's basic services and the commercial-free premium services.

Even some premium services carry commercials. Pay sports services have done so from the beginning, and periodically, there is a flurry of speculation that other premium services will follow suit. The Playboy Channel's president, Paul Klein, said in March 1983 that the channel "is interested in" selling advertising once it has acquired between 1 million and 1.2 million subscribers.

Still, the misunderstanding that all cable television is commercial-free is pervasive. In New Haven, Connecticut, a recent radio talk show featuring the head of the local cable system was peppered with complaints from viewers who "always thought that cable meant no commercials." New Haven has been wired for a number of years, so residents are not unfamiliar with cable. Nevertheless, the myth that cable means commercial free persists.

From the point of view of program producers, the need for commercials is clear. Unlike premium services, basic services

Scenes from "Volunteers," an Anheuser-Busch, Budweiser commercial. (Photo: Courtesy of Anheuser-Busch, Inc. and D'Arcy-
MacManus & Masius, Inc.)

such as ARTS and, previously, CBS Cable, do not receive subscription fees; in fact, many of these services are provided free to the cable operator. Programming must be financed by ad sales and by subsequent sales of the programming to other markets — foreign, video disc, and cassette, for example. Without attracting meaningful support from the advertising community, such services cannot survive.

Ad Agencies Are Not Sold on Narrowcasting

So far, most advertisers and advertising agencies have been wary about committing substantial revenues to cable. There are many reasons for this, one of which derives from a comparison between broadcasting and narrowcasting. For an advertiser buying time on all three broadcast networks, for example, it is relatively safe to assume that if enough time is bought, the commercial will be seen by nearly everyone who watches television. In contrast, many cable services reach very narrow, specific audiences, a fact which some advertisers consider an advantage. Charles B. Fruit, corporate media director for Anheuser-Busch, Inc., explains, "Our consumer research indicates heavy beer drinkers watch a disproportionately large amount of sports." Because of that, Anheuser-Busch began to advertise on cable, where large concentrations of sports fans could easily be found. Indeed, the strongest attraction of cable services from an advertiser's point of view is their appeal to a definable audience.

Cable's most unappealing aspect is the lack of any generally accepted methods of measuring cable's audience size. Advertisers who buy network television time have access to statistics that are accepted industrywide as accurate measures of audience size. The cable industry has not yet devised a way to measure the audience and, consequently, has no really authoritative figures to offer.

Although by the close of 1982 approximately 51 percent of all television households in the United States could receive cable and 33 percent had actually subscribed, the potential cable audience is considerably smaller than the universe of all television households. All the advertiser can be sure of, then, is that the maximum potential audience for any cable channel would be only 33 percent of the television universe. An advertising executive for a major U.S. packaged-goods firm commented on this limited universe when asked about cable advertising for his firm. "Why is it so great that for $200 you can reach one-tenth of one percent of the people? Why do you care? The cable mavens say, 'Because it's the new medium, and you gotta go with the new media,' but I don't buy it."

Audience Research: Necessary But Difficult

Traditional methods of measuring audience size pose difficulties for cable television. A.C. Nielsen uses meters in a number of homes nationwide to record when the television set is turned on and to what channel it is tuned. An obvious drawback is that the meter doesn't know whether anyone is actually watching the television, a problem for broadcast television as well as cable.

Diaries, another method of audience measurement, require the viewer to record every program watched. This method can be unreliable, since it has been shown that viewers tend to lose interest in keeping their diaries accurately and often do not record channels watched for short periods of time. This is particularly troublesome for such cable offerings as The Weather Channel, the various all-news services, and Music Television (MTV), all of which present their programming in short segments and whose viewers tend to drop in and out. Further, cable viewers are faced with the onerous task of keeping track of as many as thirty-six to fifty channels.

Traditionally, telephone surveys have been considered very accurate for broadcast television, but they can be extremely expensive to conduct. Telephone coincidentals (random calls to homes asking what program is being watched at the time the call is received) require a very large number of calls to identify patterns, and with cable, even more calls are required since there are fewer cable homes. Surveys in which viewers are asked what they have watched over a period of time are less expensive than telephone coincidentals, but they depend upon the accuracy of the viewers' memories — a dependence that may be misguided in cable's multi-channel environment.

The Cabletelevision Advertising Bureau (CAB), an industry organization which fosters advertising on cable, is addressing some of these difficulties. Together, the CAB and the National Cable Television Association (NCTA) have undertaken a major methodology study to evaluate different ways of measuring cable audiences, a study which will eventually result in a proposal of standards to be adopted nationwide.

Another problem that cable services and operators have had in attracting advertising revenues has been the advertisers' perception of the cable audience as primarily rural. Although cable began as a glorified antenna service for rural areas, the skew is changing as more and more suburban and urban centers are wired. But the old image of the cable viewer as rural persists in some quarters.

Audience research is further complicated by recent

changes in television viewing habits. Today's television audience is no longer the mass of passive viewers who, in the
fifties, went home every Wednesday night to watch Milton
Berle. Today's viewer, hooked into a cable system which may
offer thirty or more channels—including commercial-free
pay movie services—can sit comfortably in his easy chair,
flipping through all the available fare using a remote channel
switcher before settling on one choice. If a commercial is an-
noying, the viewer can zap it out by turning to another station to catch a moment of the ballet or a snippet of news or
weather. Researching the television audience was relatively
easy when there were only three or four stations available.
Cable has changed all that, and researchers are scrambling to
catch up with the changes in behavior that have resulted.

Cable's Pluses for the Advertiser

Despite the complications of researching the audience and
the relative smallness of cable's universe, cable is beginning to
look more attractive to advertisers. As audience research increases and becomes more sophisticated, one of the most attractive attributes of the cable audience is that it tends to be
upscale. Viewers have higher-than-average incomes, more
education, and hold higher-level jobs than the average television viewer. Also, the narrowcast nature of cable programming gives the advertiser a very clearly defined environment
in which to place commercials. Beer companies can target audiences they know are receptive. Likewise, cultural programming creates a sophisticated environment which is uniquely
appropriate to certain products and services. An advertiser
who knows, for example, that a large percentage of current
customers watch PBS programming can reach similar prospective customers via a cultural channel. Such a buy could
be more cost-efficient than a network buy.

Cable television also offers opportunities for advertisers to
experiment with unusual ad formats, to match their product
to the nature of the programming, and to promote themselves
for less money overall than they would have to spend for a
network campaign. Some advertisers have used cable for *infomercials*, commercials which are longer and contain more information than standard commercials. Some have commercials specially designed to appeal to a particular audience.
Other cable advertisers have been able to place their advertising near programming that is a particularly good match
editorially (Kenl-Ration sponsors "All About Pets" on Cable
News Network (CNN) and considers this a good setting in
which to promote dog food) or to sponsor entire shows, as

Kraft did the Kraft Music Hall on CBS Cable.

Although cost considerations forced Kraft to abandon its network sponsorship of the Kraft Music Hall on NBC, the show became economically feasible again with the advent of CBS Cable because the cost of sponsorship was lower on cable. Similarly, Hallmark sponsored "Kaleidoscope," a series of children's programs on the USA Network, not to sell cards to children, but as a part of its corporate commitment to young people. Anheuser-Busch's Charles Fruit also speaks of his company's ability to "direct separate campaigns to different segments of our target audience without offending or turning off others. We've been able to tailor-make creative executions to groups."

Costs for broadcast advertising are figured in terms of cost per thousand (CPM), or how much it costs to reach each thousand viewers. The number of viewers expected to see any given commercial is based on the Nielsen and Arbitron ratings. Cost per thousand is calculated by dividing the network's charge for commercial air time by the projected audience in thousands. Although there is a good deal of debate over the accuracy of the methods employed by the rating services to calculate audience size, the estimates are accepted as an industry standard. As long as the cable industry has no such standards, cost comparisons in terms of CPM's are

The cast and crew during a taping of Hallmark Cards Kaleidoscope series. (Photo: Bill Gilbert, Courtesy of Hallmark Cards.)

suspect. Absolute costs per thirty-second commercial are possible, however, and here cable is for the most part significantly less expensive than network time. According to *The New York Times*, the average cost for a thirty-second spot on "Sixty Minutes" was approximately $200,000 during the 1982–83 season. On ARTS and CBS Cable, costs for thirty-second commercials have ranged from $500 to $1,500. The absolute cost differential is impressive, but it is essential to bear in mind that the size of audiences reached is in no way comparable.

CAB President Robert Alter sums up the dilemma of ad-supported cable services neatly: "Up to now, ad agencies have always perceived television as a mass, unidimensional medium which could be dealt with via a cryptic shorthand like CPM. Agencies have never had to deal with television in terms of a fragmented audience, and that is what is developing now." Once the agencies learn to deal with the fragmented audience, Alter contends, ad-supported culture on cable will begin to have a chance.

In the past, advertisers worked on the theory that it was possible to reach anyone who was a cultural viewer at some *other* time — when watching "Sixty Minutes" or "M*A*S*H," for example. But with the increasing fragmentation of the television audience and the decline in network shares, there are segments of the population the advertiser cannot reach via network ads. Once the cable industry is able to present hard audience data, with not only audience numbers but also information on the perceived quality of programming, Alter believes there will be a real commitment to cable on the part of the advertising community.

In an interesting twist on Alter's theory, *Reader's Digest* has set out in hot pursuit of advertisers who are losing portions of their once safe network audiences to cable. In early 1983, the Digest launched an audacious print campaign, drawing attention to some startling numbers from Tulsa, Oklahoma. Citing a study done by A.C. Nielsen, *Reader's Digest* asserts that in Tulsa's cabled households, the viewers' movement away from the networks is dramatic: In homes without cable, the networks get 90 percent of the prime time television watchers; in homes wired with twelve-channel systems, the networks get only 74 percent of prime time viewers; and among viewers with thirty-six channels to choose from, only 56 percent stayed with the networks.

While all this network audience erosion is taking place, apparently magazine readership in Tulsa has remained the same in cabled homes as in non-cabled homes. There has been no erosion in reading habits, the Digest argues, citing

two studies that buttress this point. So it is clear how the
Digest wants advertisers to cope with scattered viewers. "You
can't reach this lost audience with more network TV. (And
with network costs soaring, who can afford to try?),," says one
full-page Digest ad in *The New York Times*. "You can't do it
with cable, not when cable requires a horrendous scatter-shot
buy. And you can't possibly do it with commercial-free pay
TV. You can do it with magazines. Led by the Digest — with
its astonishing 35 percent coverage of cable TV homes —
magazines offer the most effective way to reach the network
defectors."

The Digest contention is, of course, open to debate. What
is not open to debate is that *Reader's Digest* has fired the opening shot in what promises to be a long, complex battle for
advertisers' dollars. Never again will advertising be as simple
as a straight network buy.

Advertising on Cultural Cable

In the summer of 1981, *Cablevision*, an industry magazine,
sampled ad agency attitudes toward program services selling
advertising time. The comments about the cultural channels
were not enthusiastic. Advertising executives charged that
these services would be unable to provide large enough audiences to justify buying ad time. Despite the fact that part of
cable's appeal for the advertiser is supposed to be its ability to
deliver specific audiences and the widely-held belief that the
cultural audience would be very upscale (therefore have more
disposable income), many agency people contended that
cultural programming would never succeed in reaching a
substantial audience. One ad man went further, saying that
even though the ARTS studies had indicated that there was a
demand for cultural programming, the public had never supported public television and would not flock to cultural channels either.

The original plan for ARTS called for very limited advertising. ARTS' agreement with Warner Amex, which permitted
the sharing of a transponder with Nickelodeon, required that
ARTS not run traditional advertising. Nickelodeon is a commercial-free children's program service marketed by Warner
Amex Satellite Entertainment Company (WASSEC). The
match between Nickelodeon and ARTS was a good one; the
company felt that the same families who wanted their
children to have high-quality, commercial-free programs
would be attracted to high-quality visual and performing arts
programming. WASSEC, however, felt it would be inap-
propriate for the arts programming to be cluttered by tradi-

Ordinary dry dog food

Ken-L Ration Tender Chunks

Scenes from a Ken-L Ration Tender Chunks commercial. (Photo:
Courtesy of the Quaker Oats Company and J. Walter Thompson
Company.)

tional commercials, so an alternate plan was developed to find commercial sponsorships for ARTS. ARTS would locate corporate underwriters who would have limited time for billboards and infomercials. This plan attracted no buyers, so with Warner Amex's agreement, ARTS changed sales strategies and began selling thirty- and sixty-second spots and program sponsorships. Not until they made that change was ARTS able to attract any advertising revenue. Among the early advertisers on ARTS were General Motors, Ford Motor Company, Polaroid, AT&T, Mobil, and Eastman Kodak.

Although lack of advertising was cited as the primary reason for closing down CBS Cable, that service had some of the best luck in attracting advertisers who were willing to experiment with different commercial formats and with tailor-made placements adjacent to related material. The opening night of CBS Cable can provide an example of each. Certain commercials were placed in what the agency and company felt were particularly hospitable program surroundings; a dog food commercial was run adjacent to a feature on raising horses. Nonstandard-length commercials were run that night near appropriate content. CBS Cable was also able to attract Kraft to sponsor the rebirth of the Kraft Music Hall and Exxon as the sponsor of the Bernstein Conducts Beethoven series.

James A. Joyella, CBS Cable's vice president of sales, candidly admitted in the summer of 1982 that the industry rumors about CBS Cable's financial difficulties had made it increasingly difficult to sell advertising time. Advertisers do not want to commit funds to a service that may not continue to exist. A mere ten days before the announcement of the cancellation of the service, however, CBS Cable had announced new advertising commitments from Quaker Oats, AT&T, and Saks Fifth Avenue.

The overwhelming criticism voiced in the advertising community of both ARTS and CBS Cable as advertising vehicles has been that they do not have adequate audience measurement. At the time of its demise, CBS Cable spoke of 5 million subscribers and ARTS of 7.5 million. What that actually meant was that there were that many homes that could *receive* the services. There were no methods in place yet to document how many of those homes actually *watched* ARTS and CBS Cable, or how often.

If lack of definitive audience figures is the primary criticism ad agencies level at the cultural services, certainly CBS Cable staff members could criticize their corporate parent, CBS Inc., for not giving them enough time to prove themselves as an advertising medium. Ad agencies are institutions which move and change as slowly as do all institutions. Cable, as Robert Alter points out, requires changes in the current approach to advertising. To advertise on cable requires much more work; the agency has to do more research on more different aspects of the audience than is done on broadcast. When CBS Cable was axed, the agencies had not had even two full years of learning to deal with and exploit cultural cable as an advertising medium.

CBS Cable wooed advertisers by pointing out that on CBS Cable (and the same would undoubtedly hold true for ARTS), the advertiser could reach *The New Yorker* reader with video advertising for the first time. CBS Cable contended that it was attracting people who were ordinarily unreachable to television advertisers, the light viewers and the non-viewers. Their argument had a certain ring of good sense about it, but few hard figures to back it up. CBS Cable's James A. Joyella spoke of narrowing the leap of faith required of advertisers on his service. That gap was narrowing as the service closed down, but proving that an audience is one the advertiser has never been able to reach before is time-consuming.

Both ARTS and CBS Cable premiered with the hardest ad sell of any of the ad-supported networks, and their debuts came during a period of severe economic strain. The bad economy of the early eighties has not created an hospitable atmosphere for advertiser experimentation. Budgets are tight and getting tighter. It is not a time for experimentation or image building, as hard-talking ad executives like to point out.

The packaged goods executive, quoted earlier speaking against advertising on cable for his company, was enthusiastic about the efficacy of cultural cable for certain other advertisers: "Narrowcasting does provide a good strategy for some companies, and I'm surprised that some of those are not in cable right now. I think of the high-ticket items — Tiffany, for example — I don't know much about their marketing strategy, but for an advertiser who doesn't care about reaching everybody, narrowcasting is a fine way to reach a select audience with video. They do it with magazines, and now they can do it with video."

The question of cultural programming's survival as advertiser-supported programming turns on just that issue: Will cable emerge as a narrowcast medium comparable to a magazine advertising environment? Thus far, most advertisers continue to insist on *numbers*. Very few seem to care what *kind* of audience makes up those numbers.

But some do. Pet food manufacturers want to reach pet owners and record promoters want to reach teenagers. To date, however, there has been very little indication that advertisers want to reach the cultural audience. As a matter of fact, there is some indication that, despite the findings and projections of CBS Cable and ARTS, the cultural audience may not be as clearly identifiable and relatively monolithic as was once thought. Thus, the survival of ad-supported culture will depend on several things: the ability of cable to demonstrate that it is reaching significant numbers, a clearer and more convincing characterization of that audience's identity, and advertiser acceptance of cultural cable as a viable narrowcast medium.

Culture by Satellite

In April 1981, there were three satellite-delivered program services devoted entirely to culture, two actually in operation and one close to launch. There were also three general audience channels featuring some cultural programming.

Two years later, only two of the cultural services remained, and one of those had become 60 percent movies and 40 percent performing arts. Of the general audience channels, only one continued to offer any cultural programming.

What happened in those two years? Is this truly the end of the second golden age? Or are the Madison Avenue people right when they say that culture and the arts have never attracted significant audiences and, consequently, cannot survive on television? For all the extraordinary excitement created by announcements of cultural programming ready to enrich the vast wasteland, there seems to be disappointment everywhere. A brief history of each of these services will help to explain the reasons for disappointment and to indicate the prospects for survival of culture on national cable services.

ARTS

ARTS, the only surviving exclusively cultural network, is a joint venture of American Broadcasting Companies, Inc. (ABC) and the Hearst Corporation. Launched in April 1981, ARTS offers programming which covers "the full spectrum of expression in all the performing and visual arts, from the most traditional to the most contemporary, from the performance through the documentary, from the economics of arts to the sociology of arts. It also deals with aesthetic issues," according to Mary Ann Tighe, who functioned as the service's chief programmer during 1981 and 1982.

ABC's 1981 annual report contains a clear statement of the company's strategies regarding the new television technologies. First is ABC's commitment to "supply the broadcast operations with the resources necessary for continued success in [the] evolving communications environment; second, to move quickly into the creation of new program services suitable for the new media; third, to explore other opportunities that may become available as a result of future changes in regulation of technology." Because ABC considers its strength to be in programming, not technology, the corporation moved rapidly into cable programming, primarily through joint ventures with other companies. The joint venture with Hearst Corporation has thus far produced two program services, ARTS and Daytime, a service created for women. ARTS, the annual report explains, "is based on the premise that while cultural events cannot draw a huge paying audience on television, the viewers who will watch but not pay are numerous enough to appeal to advertisers; with a controlled program budget, this service should be moderate in size and attractive in results." This is a modest picture compared to the report's description of ABC's cable activities in sports and news as "exceptionally appealing," and the description of the "immense potential" of Daytime for advertisers.

ABC decided, in 1979, to enter the new television technologies. More than a year and over a million dollars were invested in researching precisely what directions would be most promising. The arts were chosen for three reasons. First, as a relatively small business that could rely initially on abundant and affordable European programming, ARTS would not require a huge capital outlay. Second, the Public Broadcasting System (PBS) had demonstrated that it was possible to secure corporate sponsors for arts programming. Third, the projected "upscale" audience that would be attracted by the arts programming was prestigious and different from the network audience. Consequently, an ABC entry into arts programming would not offend the broadcast affiliates by competing with them for audiences.

When Nickelodeon's transponder became available to ABC, the whole deal fell into place. Not only did the pairing of a high quality children's service during the daytime with a performing and visual arts service in the evening seem ideal, but sharing a transponder with Warner-Amex's already-established Nickelodeon enabled ARTS to begin with an assured number of subscribers (3.5 million) when it was launched, a number that grew to 9.5 million by the end of its second year.

Part of the original agreement between Hearst/ABC and Warner Amex required that ARTS run very little advertising. The original plan, in fact, called for seven corporate underwriters using only "infomercials" and billboards to display company logos. Unable to attract underwriters, ABC was forced to renegotiate the deal with Warner Amex so that ARTS could sell time to advertisers in traditional thirty- and sixty-second spot form. Although not fully ad-supported during its first years, ARTS has attracted advertisers.

The first year's programming was primarily European imports, bought relatively inexpensively before the demand for cultural programming skyrocketed. Approximately 85 percent of the first year's programming was acquired, not all of it from overseas, but enough to prompt *New York Times* television critic John J. O'Connor to comment, in March 1982, that "ABC cannot coast on old European productions forever. . . ."

ARTS debuted with theme weeks as a unifying concept for the programming. The first three weeks were: "Paris: The Dream and the Reality," hosted by Anne Baxter; "Vienna: The Home of Genius," with Pierre Salinger; and "Paris: The Artist's Heaven," which included performances of Book I of Debussy Preludes, a documentary on Edgar Degas, and a half-hour version of Ravel's "La Valse," directed and choreographed by George Balanchine and danced by members of the New York City Ballet.

Some first-year viewers complained bitterly that there was no way to figure out the scheduling, so in place of the somewhat artificial device of theme weeks, a new format was introduced with considerable fanfare at the beginning of the second year. On Sundays were opera, classical music, and visual arts documentaries. Mondays featured dance and jazz. Tuesdays were reserved for theatre, comedy, and drama, and on Wednesdays there were Mobil Showcases or "encore presentations" of previously shown programs. On Thursdays, Fridays, and Saturdays, the Sunday, Monday, and Tuesday programs were repeated.

Yet another evolution of scheduling took place shortly after the third year began in 1983. What had been a clear repeat pattern was broken for several reasons, according to Curtis Davis, ARTS' director of programming: to add more variety, to schedule repeats on evenings when the fewest viewers watch the channel (weekends), and to make viewers less aware of how often programming is repeated.

But the changes in schedule were minor compared to the changes in programming during the second year. Gone were the theme weeks with American hosts. Instead, ARTS programming was allowed to speak for itself. The image of the service was spruced up with new electronic graphics. But most important was the new programming created for the channel at the behest of ABC Video Enterprises' vice president for program development, Mary Anne Tighe. Tighe arrived at ABC Video Enterprises in summer 1981, several months after the ARTS service was launched, and took over the responsibilities for program development. "I felt that

Hearst/ABC's ARTS production of David Mamet's play *Dark Pony*, starring Lindsay Crouse and Michael Higgins. (Photo: Courtesy of ABC Video Enterprises.)

there should be a commitment to more contemporary art as well as the traditional arts," she said. "I think ARTS must always have a balance of the traditional and the contemporary. But, I like the idea of using particularly American theatres and institutions. I want to show not just New York and Los Angeles, but any place where interesting work is being done."

Tighe proceeded to re-shape ARTS' identity by adding work produced or co-produced by ABC Video Enterprises. New American drama was presented: David Mamet's *Reunion* and *Dark Pony* and Frank South's *Rattlesnake in a Cooler* and *Precious Blood*. A production agreement was signed with Joseph Papp to produce six hours of programming for ABC Video Enterprises. A joint production agreement with the ABC News Documentary Unit produced the *Eighteenth Century Woman* and *La Belle Epoque*, based on Metropolitan Museum costume exhibits.

Dance documentaries included *A Portrait of Giselle*, nominated for an Oscar in the Feature Documentary category in 1982, *To Dance for Gold*, which used sports camera techniques on the Jackson, Mississippi International Ballet competition performances. Commentary for that documentary was provided by Jacques d'Amboise, Dick Button, and Marge Champion. The footage which resulted from the use of the sports techniques was exceedingly engaging, with Jacques d'Amboise providing just the proper level of commentary for the interested, but unsophisticated, viewer.

Moses Pendleton Presents Moses Pendleton is a stunning performance documentary discussed later in this book. On the other hand, straight performance footage of the Alvin Ailey American Dance Theater, taped live in performance at New York's City Center was a disappointment.

Music programming created by ABC Video Enterprises has included a one-hour special on Beethoven's life, conceived by pianist Israela Margalit, and a series entitled "Women in Jazz."

ARTS' programming has varied widely in its quality. Because so much of the programming is acquired, the "look" of the channel is not consistent as it was with CBS Cable. But the best quality of ARTS is very high. The variety of programming offered by ARTS is impressive — from visual arts documentaries to a series on craftsmen, "Handmade in America," and a series called "At the Met: Curators Choices." ARTS has also been adventurous in its programming. Two examples are *Swan Lake: Minnesota*, a dramatic variation on the theme of the ballet *Swan Lake,* and Robert Wilson's video piece *Stations*, both outstanding productions that demonstrate that the channel's programming policy is not guided by cautious or conservative heads. ARTS has licensed some fine material, including Harold Pinter's *The Collection*, with Alan Bates, Malcolm McDowell, Laurence Olivier, and Hellen Mirren. This screenplay is one of the finest transfers of a play to the screen ever accomplished. Despite its uneven quality, ARTS has provided some extraordinary evenings of programming to viewers around the country.

Will ARTS survive? The fact that it still remains while other similar services have folded is in its favor. Yet, the service is still losing money, and can not firmly predict when it will break even.

In the beginning of 1983, ARTS (which had earlier been administered by ABC Video Enterprises for the Hearst/ABC partnership) was moved entirely under the management of the Hearst/ABC unit which also produces Daytime. In the process, the service got a new vice president for programming, Mary Alice Dwyer-Dobbin, although Curtis Davis, the programming director for ARTS "since day two," remained. What impact Dwyer-Dobbin will have on the service remains to be seen.

Although there has been talk that ARTS may not survive 1983, in an April 1983 interview, Herb Granath, president of ABC Video Enterprises and co-chairman of Hearst/ABC Service's board of directors, spoke about ARTS' prospects. "We are still losing money on ARTS, and we anticipated losing money on it for three to five years when I took the plan to

(Above and opposite page) Hearst/ABC's ARTS production of "Joseph Papp Presents: Swan Lake Minnesota," an innovative adaptation of *Swan Lake* about a farmer who falls in love with a ballerina who dances in and out of his mundane life. (Photos: Dale Wittner, courtesy of ABC Video Enterprises.)

the (ABC) board. What we have to do as managers of this service is to maintain the losses at levels that are acceptable, and *so* far we have been able to do that."

ARTS expects its income to grow as its subscriber numbers grow. "The advertising community," said Granath, "indicates that critical audience mass is somewhere between 10 million and 12 million subscribers. We are at 9.5 million now and expect to be somewhere near 12 million by year's end."

Further, Granath expects subscriber numbers to grow almost automatically, with little necessity for ARTS to invest additional funds in that growth. The reason for this is that "virtually every new franchise bid included ARTS, because it was 'the thing to do.' " So as the new systems are turned on, subscriber numbers will increase. Granath also forecasts growth from increased channel capacity that he predicts will result from older systems rebuilding to renew their expiring franchises. "There is not *one* of those older systems—some 40 percent of the nation's cable systems with only twelve channels—that will be allowed to continue as a twelve-channel system," he said. "They will all have to upgrade to at least thirty-six [channels], and if you go to thirty-six, the in-

cremental cost of going to fifty-four is negligible. Consequently, the channel capacity problem we have had will not continue to be a problem as time goes on, meaning within the next year and a half."

Granath is certainly right that there will be virtually automatic growth in newly constructed systems. His estimate of the prospects for increased channel capacity in older systems is highly optimistic. Many in the cable industry would dispute both his optimistic timetable and his assumption that the nation's twelve-channel systems are so certainly headed for an increase to thirty-six, or even fifty-four, channels in the near future. (The industry trade association, the National Cable Television Association, is working hard for the passage of national legislation that would grant virtually automatic renewal of franchises nationwide. Passage of such legislation could seriously curtail the prospects for ARTS' growth in older cable systems.)

Nevertheless, if ARTS can pick up another 2.5 million subscribers, it will have entered the area of "critical mass" for advertisers. That subscriber number, however, represents *potential* viewership, not necessarily actual viewers. Adver-

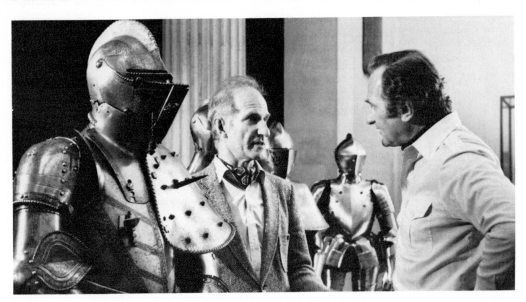

A scene from Hearst/ABC's ARTS production of "At the Met: The Tournament," the third in a series based on the collections and resources of The Metropolitan Museum. (Photo: Courtesy of ABC Video Enterprises.)

tisers not only want the ten to twelve million critical mass figure, they also want some demonstration that people are *actually watching* the service. ARTS' viewership has been monitored by Nielsen, but the figures have not been released. Demonstration of actual viewership must be the next step for ARTS' ad sales.

Thus, the question of survival remains inconclusive. Granath commented, "I think I could say that if our losses mounted at an alarming rate in our third year, going into our fourth year, over what they were in the preceeding years, then we would certainly, as good businessmen, have to look at the situation and say we made some miscalculations. However, we have some control over that. We have some control over ad revenues and we can control the costs of the service."

On the plus side, the cost of acquiring programming has declined because of the deaths of both CBS Cable and The Entertainment Channel. There are simply fewer buyers now. The hopes for an upturn in the U.S. economy also provide hope for an increase in ad revenues as well. Finally, if the service makes it into 1984, there will be considerably more money available for non-network advertising. "Two Olympics and a presidential election put a very heavy squeeze on network television ad availabilities," explained Granath. "Therefore, there should be a considerable amount of money spent in the alternative media—hopefully, in cable."

Given these facts, it is rash to predict the outcome of the ARTS venture. As it enters its third year, it remains a modestly funded service trying to carve out a place for itself with the viewer, the advertiser, and the cable operator.

Bravo

Bravo, a pay service featuring a mix of international and critically acclaimed U.S. films, along with performing arts programming, currently has approximately 101,700 subscribers in 71 different cable systems. Programming runs ten hours a day, seven days a week, from 8:00 P.M. to 6:00 A.M. EST.

Although Bravo now cablecasts a mix of film and performing arts programs seven nights a week, when the service began in December 1980, it provided only two nights of cultural programming and shared the rest of the week with Escapade, an R-rated blue-movie service. This arrangement continued until July 1981, when Bravo separated from Escapade and began adding American and foreign films to its program mix to fill out its present full-week schedule.

Bravo was created by Rainbow Program Service, a joint venture of three MSOs: Cablevision Systems, Cox Cable Communications, Inc., and Daniels and Associates, Inc. Many speculated about Rainbow's motives in initially offering culture and blue movies to subscribers in the same package. Rainbow sources explained at the time that only one transponder was available, and that pairing an adult movie service with culture would help the cultural service financially. In other words, popular adult fare would subsidize the arts. Additionally, cable system operators made squeamish by the objections of Moral Majority people to adult services like Escapade could offer Bravo as a moral sweetener. Strange business partners, observers murmured, and cynicism about Rainbow's real commitment to the cultural side of the service abounded. But cynics who saw Rainbow using culture solely as a Trojan horse for R- and X-rated movies have been proven wrong; Bravo has survived its separation from Escapade.

Bravo was created to meet an established demand for cultural programming, according to the service's first president, Marc Lustgarten. "We asked our Cablevision subscribers what they wanted, and cultural programming kept coming up in the responses." Rainbow realized further, according to Lustgarten, that it was useless to create just another movie service. "There were already too many movie services showing the same films." Instead, the organization sought to create the kind of service that would compliment existing ones and attract additional subscriber fees. The decision to make Bravo a pay service was based upon Rainbow's firm conviction that culture would never attract audiences large enough to convince advertisers to support the channel.

The current programming mix on Bravo is 60 percent

The Canadian Brass Quintet on
the set of "Bravo Magazine," a
Rainbow Program Service. (Photo:
Charles Abbott.)

movies and 40 percent performing arts. A trade advertise-
ment run in April 1983 explained it this way: "So what is
Bravo? Arts? Or movies? We're both. Forty percent performing arts, 60 percent movies. And we're by no means esoteric.
We're a cultural service, with an enormously broad appeal."

The performing arts material shown on the channel is
heavily weighted with musical events: opera, symphonic
music, and jazz. Nearly 70 percent of the performing arts
programming is musical. This is partly due to a pioneering
agreement Bravo negotiated with the American Federation of
Musicians (AFM) at the very beginning of the channel's production life. Lustgarten, in 1981, stated that the AFM agreement "has allowed us to work on a more economical basis. We
produce in-house in a highly efficient way, and can produce
complete symphonic music for one-third of what Channel 13
spends." Bravo, however, has not been producing any of its
own programming since March 1982, but the stress on music
programming persists.

From April 1982 to April 1983, the orchestras that per-
formed on Bravo included the St. Louis Symphony, The
Cleveland Orchestra, the Baltimore Symphony and the
American Symphony Orchestra. The complete Brandenburg
Concerti were taped at New York City's 92nd Street Y, and
Allan Miller (who was artistic supervisor for the film *From
Mao to Mozart*) produced and directed a documentary on the
1981 International American Music Competition for Bravo.
The documentary follows twelve competitors through the
days that precede the competition and includes a substantial
amount of performance footage as well.

Music for Wilderness Lake, another Bravo offering, is an

unusual documentary on environmental music. This film traces the development and performance of a piece of environmental music in which twelve trombonists station themselves around the shore of an unpopulated lake. The musicians are conducted by composer R. Murray Shafer, who is positioned on a raft in the middle of the lake.

Included among the operas that Bravo has cablecast are two performed by the Indiana University Opera Theatre, *The Greek Passion* and Leos Janacek's seldom-performed *Excursion of Mr. Broucek*. Both were taped by Bravo. Another rarely performed work shown on Bravo is George Gershwin's lyric one-act opera, *Blue Monday*, which is one of Bravo's European acquisitions.

Jazz programming includes a series of four evenings taped at The Station in Wilkes Barre, Pennsylvania, a converted railroad station now functioning as a restaurant and concert hall. Jazz masters taped there include Dizzie Gillespie, Gerry Mulligan, Herbie Mann, and Dave Brubeck. Two other jazz programs include Bobby Short, taped at the Cafe Carlyle in New York City, and the actor Dudley Moore performing with his trio.

Dance seen on Bravo has included Ballet West, taped by Bravo in Salt Lake City; the Joyce Trisler Dansecompany; Pilobolus; a french-made *La Sylphide* with Ghislaine Thesmar and Michael Denard; and Rudolph Nureyev's *I Am A Dancer*.

The Indiana University Opera Company's production of *The Greek Passion*, taped live by Bravo. (Photo: Dave Repp.)

The performing art form least in evidence on Bravo has been theatre. While all the other cable services were caught up in the theatre on television boom, Bravo stood on the sidelines. Finally, in its third year, Bravo began to show theatrical programming, scheduling *Tintypes, Pippin,* and *Kiss Me Petruchio* for that year.

The films offered by Bravo constitute a veritable feast for film buffs. Since Bravo began programming films, a number of festivals have been presented. Included were a Woody Allen Festival; four films by François Truffaut; Simone Signoret films; seven films from Australia including *The Getting of Wisdom, Breaker Morant, Picnic at Hanging Rock*; three Federico Fellini films; a selection of German films including *The Tin Drum, The Marriage of Maria Braun,* and *Nosferatu: The Vampyre.* Stanley Kubrick's work has been featured as well as that of a number of independent filmmakers. *Days of Heaven* was cablecast by Bravo, and in March 1983, an Academy Award Festival featured winners from previous years.

Bravo has not had an easy time finding its market. Sharing a transponder and its nightly identity with an adult program service in its early months of existence was very confusing to the viewer. When Bravo expanded its programming schedule from two to seven nights a week and shed Escapade, becoming a culture *and* movie service, its subscriber roles hardly grew at all. After a year, a Bravo marketing executive candidly admitted, "We are not quite sure where we stand in the marketplace." With the addition of movies, it was hard for observers to tell whether Bravo was a movie service or a classy cultural service with art films.

Slowly, Bravo evolved an identity of critically acclaimed films with performing arts. A Bravo executive explained that it had become a service for "people who like some of the movies on HBO and really want to watch more and better movies while receiving some exposure to the arts. Bravo is not for the elite. You don't have to be a big fan of the arts to enjoy what we have on Bravo. It's really a supplementary movie service filled out by arts programming." As this book goes to press, that description still applies to Bravo.

Perhaps the most interesting development for Bravo, however, has been the latest refinement in its marketing strategy. Bravo had been marketed as a single service when it separated from Escapade, with disappointing results. The latest plan devised by the company has apparently been strikingly successful in the Cablevision, Long Island system owned by one of Bravo's parent companies.

Rainbow now suggests that operators not sell Bravo as a single service, but as a "partner" to other pay services. It is no

longer promoted for its value *per se*, but for the value it adds to a package.

This strategy is based in part on research done by Rainbow and others. That research supports the thesis that many cable viewers find cultural programming attractive, but do not want to pay very much extra for it. Since many cable systems offer several pay tiers: one with HBO or Showtime, a second with a supplementary (non-duplicating) movie service, and perhaps a third with a separate pay sports channel, Bravo would have to be in the fourth tier before it could be added to such a system's offerings.

The costs begin to mount quickly. Using easy, round numbers of ten dollars for basic service, an additional ten dollars each for the two movie tiers, and eight dollars for the sports tier, the family that also wants to add Bravo as a fourth tier will already be paying thirty-eight dollars per month for cable television. To add Bravo would cost at least an extra five to eight dollars, bringing the total bill to as much as forty-six dollars a month.

Bravo's marketing people want cable operators to use an entirely different configuration. They suggest a scheme in which the subscriber can have: basic service for ten dollars a month; tier one, offering basic plus HBO, for eighteen dollars; tier two, offering basic, HBO, and Showtime, for twenty-six dollars; and the next tier (the top) offering basic, HBO, Showtime, Bravo, The Playboy Channel *or* The

Jazz trumpeter Dizzy Gillespie in a concert taped by Bravo at "The Station" in Wilkes-Barre, Pennsylvania. (Photo: Courtesy of Bravo, a Rainbow Service.)

Disney Channel, and Sports Channel *or* Cinemax, at thirty-nine dollars per month.

The success of this sort of marketing has reportedly been remarkable. "When Bravo is put in a package with other services, we tend to at least double our audience. Sometimes it triples," said Bravo's marketing director, James Forbush. "We have had people say to us, 'Yes, when I first got cable I watched a lot of HBO and Showtime — all those blockbusters they advertise — and it was great! The third month, they repeated all those films. So I started flipping the dial and found Bravo. Now I watch Bravo every week.' Some people went further, saying, 'If you tried to sell me culture separately, I might not have bought it. But the fact is that when I'm tired of watching *Smokey and the Bandit*, and I want something more challenging, I turn to Bravo. I like it.' "

Apparently, having Bravo in a package with other pay services tends to keep customers happy (consequently paying their bills) by offering them something different. There is very little duplication between Bravo and other services, so there is usually something new to watch. And Bravo's cost to operators has been kept low, so they can package Bravo with other pay services.

If this kind of marketing continues to succeed as it has in Cablevision's systems, there is hope that Bravo will survive as a pay service that offers some cultural programming.

Ovation

Begun in 1979, The English Channel (now called Ovation) was the first cultural service on cable television. The idea came from a most unlikely source, the Edward E. Finch Company, an organization that merchandises the prizes given away on game shows. The company had recently syndicated a golf tournament to a series of cable operators and, as a result of that experience, decided to put together other programming to appeal to the educated, thoughtful, and affluent audiences to whom, they assumed, most network fare did not appeal. The Finch people approached Granada Television of Britain and reached an agreement to use their program library — The English Channel was born. It was originally a several-hours-per-week service that was distributed by Satellite Program Network. USA Network picked up the service in 1980 and changed the name to Ovation.

Today, the programming comes from Granada, Yorkshire Television, and the National Film Board of Canada, among other sources. Occasionally, U.S.-produced programming is used as well. Programming ranges from the arts to documentaries. In 1982–1983, the service carried the entire "Brideshead Revisited" series, giving it a second play following its PBS debut.

Other programming has included a four-part feature of Roosevelt's New Deal for the arts and a program entitled "The Light Fantastic," a Canadian National Film Board retrospective on animation. *K-2: The Savage Mountain*, a Yorkshire Television documentary, followed a 1978 British attempt to scale the famous mountain. According to a service executive, the film contains "the most breathtaking sequence of climbing at altitude ever shot." *The Secret Life of an Orchestra* detailed the private thoughts of musicians in the Denver Symphony. Another Yorkshire program examined the proliferation of nuclear weapons, and Alan Whicker presented a two-hour documentary on Bombay, India entitled *Don't Feel Guilty About Your Cadillac and I Won't Feel Guilty About My Bicycle*. "Ride On," an Australian mini-series has also been seen on Ovation.

Currently, USA Network features Ovation from 3:00 P.M. to 5:00 P.M. and 10:30 P.M. to 12:30 A.M. on Sundays and 3:00 P.M. to 5:00 P.M. (EST) on Saturday afternoons.

The Entertainment Channel

Launched in summer 1982, The Entertainment Channel (TEC) lasted only nine months before its corporate parents pulled the plug. TEC, a pay service which cost subscribers from eight to twelve dollars a month, was a joint venture of RCA and Rockefeller Center, Inc.

Angela Lansbury in The Entertainment Channel's presentation of *Sweeney Todd*. (Photo: Courtesy of RKO Videogroup.)

TEC was characterized often by the press as a cultural channel, counter to the intent of the service's joint owners who, from the beginning, seemed to have felt that the chances of cultural channels to succeed were slim. Yet, during TEC's two-year planning process, the only product with which to establish an identity for the channel was BBC material. (Arthur Taylor, the service's chairman, had negotiated exclusive U.S. rights to all BBC programming.) The BBC deal telegraphed a cultural programming message to the rest of the world, a message the channel worked hard to overcome.

Extensive market research had convinced TEC executives that a large portion of television viewers were displeased with both conventional television and the options available on pay television. Fifty-eight percent of the TEC survey respondents said that poor quality programming was the main reason for their disenchantment with television. TEC's programming mix was intended to satisfy those who voiced these objections, and was backed by a promotional campaign aimed at that au-

George Hearn being filmed by
cameramen in The Entertainment
Channel's production of *Sweeney
Todd*. (Photo: Courtesy of RKO
Videogroup.)

dience. "What Television Should Have Been in the First
Place" and "Stop Just Watching TV and Start Enjoying It"
were among TEC's marketing slogans.

According to Arnold Huberman, programming vice president, programming was aimed at people who, although
bored by the offerings of both the other premium cable services and the broadcast networks, were nonetheless uninter-
ested in the "rarified" atmosphere of the cultural channels.

Broadway plays were an important part of TEC's programming mix. When asked shortly before launch time why the
channel was so convinced that the Broadway mix would win
viewers, Huberman replied, "The audience is telling us. All
the research we get says the audience wants to see Broadway
plays. Theatre attendance has increased 150 percent in the
last ten years, and it's our belief that a good story line and

a good piece of entertainment will work on television. Additionally, you're giving people something to which they can attach a value. They *know* how much it costs to go to the theatre in their communities." Among the first Broadway shows to be announced were *Sweeney Todd, Pippin, Lena Horne: The Lady and Her Music, Piaf,* and *I Do, I Do.* Other theatrical productions planned included an adaptation of *Canterbury Tales*; Emlyn Williams as *Charles Dickens*; *The Drunkard*, starring Tom Bosley; *Candida*, starring Joanne Woodward and Jane Curtain; and the Guthrie Theater production of *A Christmas Carol.*

Despite its attempt to attract subscribers from among those likely to respond to the channel's quality image, TEC could not attract and hold enough subscribers to convince its corporate parents to continue the service. Over eight months, it was only able to attract 40,000–50,000 subscribers. And of those who did subscribe, industry sources reported that a startlingly large percentage dropped their subscriptions after a short time. Another difficulty faced by the service was its satellite which, at the time, was received by only a limited number of cable systems. Faced with losses estimated at between $30 million and $50 million, RCA and Rockefeller Center Inc. decided not to continue the service beyond March 1983. Arthur Taylor announced at the time that the service would reappear as a basic service in the future, but industry sources viewed this announcement with skepticism.

John J. O'Connor, writing in the Arts and Leisure section of *The New York Times* shortly after the announcement of TEC's closing, expressed the opinion of many who had watched the service closely. "TEC's bill of fare was highly uneven, falling somewhere between the class acts of public TV and the better situation comedies of commercial television. It is not surprising that viewers hesitated in paying a monthly premium of between $8 and $12 for such a service. . . . apart from the occasional success of a "Sweeney Todd," [the programming] tended to be surprisingly bland. Much of the BBC material, which accounted for a third of the schedule, consisted of situation comedies and adventure series that made their American counterparts look quite accomplished in comparison. The Entertainment Channel did put together a schedule that, on the whole, was better than the average network lineup. But, considering the monthly charge, it really isn't that much better."

Other Services Offering Cultural Programming

HBO and Showtime, the two leading premium services,

have supplemented their predominantly movie services with limited theatrical programming. In both cases, taped theatrical events have been added to broaden the appeal of the service and to add what is called "perceived ticket value." People know how much tickets to Broadway theatre and to their local theatre cost. In early 1983, however, HBO drastically reduced its theatrical programming, and its director of theatrical development since 1980, Arthur Whitelaw, resigned.

HBO had been programming one theatrical production each month as part of the HBO Theatre series, which began in March 1981 with *Vanities*. Plays taped by HBO include: *Sherlock Holmes* (taped at the Williamstown Theatre with the original cast of the 1981 Williamstown production), *Table Settings, Plaza Suite, Wait Until Dark, Barefoot in the Park, The Deadly Game, Bus Stop, Camelot* (the Broadway and touring production starring Richard Harris), *The Rainmaker,* and *Separate Tables.* Because the HBO audience was not used to theatre on television, there had to be a "hook" to attract the viewers to the experience, and all of the above productions reflected HBO's policy of presenting only well-known plays, or lesser-known plays with name stars.

HBO closely monitors the reaction of viewers to its programs and, in an interview, Whitelaw talked about the reactions of viewers to *Sherlock Holmes.* "Our viewers didn't like

HBO's production of *Sherlock Holmes* (top) and *Barefoot in the Park*, both part of HBO's "Standing Room Only" theatre series. (Photos: Courtesy of HBO.)

Sherlock anywhere near as much as they liked plays like *Plaza Suite* and *Barefoot*. I was very pleased with *Sherlock*, but I agree that it was dark and strange to many of our viewers."

According to Whitelaw, research eventually showed that not enough people watched and valued the theatrical programming on HBO to justify its being continued. "In some cases a lot watched — *Camelot, Barefoot in the Park, Plaza Suite, Vanities*. But there were others people didn't watch. There is a *masterpiece* on this month, *Separate Tables*, which *nobody* is watching, and it got the best reviews of anything [HBO has] ever done."

What theatre programming did do for HBO, according to Whitelaw, was give the service visibility. "They don't get the reviews or the press coverage with any other kind of programming, and that gave them visibility. But with *Camelot* costing upwards of $3 million and a made for television movie costing $2.5 million, it makes more sense to put the money into film. And that's what the viewers say they want to see. The theatre programming was their prestige programming. It gave them clout within the industry, in the press, and gave them exposure they wouldn't normally have had."

HBO's emphasis has shifted, according to Bridget Potter, vice president for original programming, from stage plays to developing more original programming projects. The first of these is *Mr. Halpern and Mr. Johnson*, an original, videotaped screenplay by Lionel Goldstein starring Laurence Olivier and Jackie Gleason. Whitelaw thinks this is probably a very interesting idea for the future. "Original programming for television is probably where all these new writers, who are developing in the theatre outlets I involved HBO with, will eventually see their work done."

Showtime was cable television's theatrical pioneer in 1979, when it established the "Broadway on Showtime" series which premiered *Bullshot Crummond, Monteith and Rand,* and *Tuscaloosa's Calling Me. . .But I'm Not Going!* during its first three months. There has been a new "Broadway on Showtime" production nearly every month since then. The range of programs featured on the series has been relatively broad, including the Peking Opera, John Curry's Ice Dancing, and The American Dance Machine. Early productions included *Look Back in Anger*, with Malcolm McDowell, directed by Lindsay Anderson; Jules Feiffer's *Hold Me*; Eugene O'Neill's *Hughie*, starring Jason Robards; and Jay Broad's political play, *A Conflict of Interest* with Barnard Hughes — all of which were somewhat more adventurous than HBO's theatrical offerings.

Early in the life of "Broadway on Showtime," there was, ac-

The cast of Showtime's presentation of Paul Osborne's play *Mornings at Seven*. (Photo: C. David Batalsky.)

cording to Caroline Winston, vice president for programming, "a small number of viewers who were very pleased by what they saw; it was justification for them to pay the monthly fee." As the Showtime audience grew, more people watched "Broadway on Showtime," and research began to show that the audiences wanted better programming. Officials of the service interpreted that to mean more "commercial" programming, and as a consequence began to stress the necessity for more recognizable stars and properties.

Robert Rubin, vice president for programming for Group W Cable productions, noted this change in spring 1981 at a New York University conference on cable and the arts: "...last year we sold a wonderful production of the Goodspeed's *Little Johnny Jones* to Showtime. It was not a star vehicle, and we sold it without a star. It was a wonderful show of which we are very proud, but I doubt that we could sell it to Showtime today, because it does not have stars and is not a commercial television clone. Unfortunately, cable seems to be moving in the direction of cloning commercial television."

Roughly a year later, Caroline Winston confirmed that the trend to stars and recognizable commercial properties continues. "And," she added, "we are also bringing in the talent of people like Robert Altman and Vivian Matalon, people whose names may not be household words, but the property you get is better—better stars, better acting. Yes, we're going after better talent in general."

Clearly, the theatre boom on pay television services is over.

Only Showtime will continue to present these theatrical productions, featuring highly commercial properties and recognizable stars.

There had seemed to be a very real demand for this, as reflected in the results of an informal telephone interview survey of cable subscribers conducted by the industry magazine *Cablevision* for its May 10, 1982 issue. One of the respondents, William Morris of Tulsa, Oklahoma, was asked what he would like to add to his cable company's offerings: "If we had to pick one item we'd like to see more of, perhaps a Broadway play. The legitimate theatre on television. We're talkin' Broadway, now, rather than opera and ballet. We have a ballet and an opera here, and every once in a while you'll see something on our educational channels that you could watch — and you don't. So I guess we would tend to watch more of a musical — *Annie, Chorus Line*, that sorta thing, rather than operas. I read *New York* magazine, and I read the reviews in there. There are some plays that are on now that are expensive to fly up for the weekend to see. If it would be possible to just set the camera there while the stage show was going on, blank out New York and surrounding areas so they aren't hurt there in their income, and let the people out here pay some money to see that, well, I'd be right pleased." It is exactly those wishes that Showtime, HBO, and The Entertainment Channel were trying to fulfill through their theatre programming. At this point, only Showtime persists.

One industry observer who hopes to see well-created theatre survive on television says HBO's theatre failed for two reasons. First, HBO didn't promote it well enough. "Viewers tend to assign value according to how much promotion an event gets, and theatre didn't receive enough promotion." Secondly, "HBO didn't create good televised theatre; it was horrible. You don't need *two* stages — the television and the proscenium — that puts the viewer too far away. When you tape directly from the stage, the actors appear to be shouting; they over-gesticulate and look mannered and artificial. As a result, viewers don't enjoy watching them." These same criticisms could be leveled at some, but by no means all, of the Showtime theatrical productions.

When asked what kind of theatre had been well done among cable services producing theatre, the same industry observer replied that The Entertainment Channel's *Sweeney Todd* was one example. "That production of *Sweeney Todd* used the stage as a *studio*, taking a massive production and achieving intimacy. It was brilliant! And it showed us that you have to re-light and re-direct any theatre that is done for television." The theatrical presentations of CBS Cable, along with British productions, were also cited as ground-breaking examples which have established standards that any serious producer of theatre for television will have to aim for in the future.

What does this all mean for culture on the national services? The Entertainment Channel found that people did not want culture and worked hard to rid itself of the cultural channel label. CBS Cable did not make it. Yet, Bravo and ARTS survive for the time being. It is too soon to be able to draw any conclusions from these ambiguous events.

"Utopia with Aggravation": The CBS Cable Story [1]

An elegantly packaged tin of caviar accompanied CBS Cable's announcement of the results from its first survey of audience demographics. "An Audience that Appreciates," proclaimed the accompanying brochure. In a busines where most premium promotional items are still at the level of the coffee mug with the company logo, the CBS caviar mailing was symbolic of the style CBS brought to cable.

Known for its elegant parties as well as its generous programming budget, CBS Cable went first class all the way from its announcement to its demise. Its quality was never compromised, and the critics appreciated this as much as the viewers did. Never has a cable service received such universal critical praise.

No doubt about it, CBS Cable arrived with a silver spoon in its mouth. How could this darling child fail so soon, before it even reached its first birthday?

The service seemed to have nearly everything going for it. CBS assembled a superior programming staff, and gave it a program budget more generous than those of other basic cable services. CBS created a first-rate marketing staff (headed by a woman whose brainchild the service had been) and offered its sales people back-up on a par with competitors in the cable industry. Because selling *cable* ads is not like selling *broadcast* television ads, CBS chose its vice president of ad sales from the radio advertising business, following the lead of the Cabletelevision Advertising Bureau.

Finally, CBS Cable had the stamp of approval from the redoubtable William Paley, the chairman and founder of CBS Inc., who had always wanted to present this kind of quality on network television. (Not only did William Paley support the new service, so did Thomas Wyman, hired to be Paley's successor while the service was in the planning stages.) In *The Powers That Be*, David Halberstam describes Paley's well-known business acumen: "No one read a business report better than Bill Paley; he had an unerring instinct for the bottom line, just as he always seemed to ask the right

---

1. Charlotte Schiff-Jones' description of her experience working at CBS Cable.

Was CBS Cable a Trojan horse created solely to get CBS Inc. into the cable business without offending its broadcast affiliates?

question at every business meeting, the one question that uncovered the fatal weakness in any proposal."

The fact that CBS Cable appeared to have nearly everything working in its favor has caused a great deal of speculation as to why, in the end, the service was killed in its infancy by the corporate parent. Given Paley's well-known business judgment, is it possible that CBS Cable was set up to fail? Was CBS Cable a Trojan horse created solely to get CBS Inc. into the cable business without offending its broadcast affiliates? Or, was the approval on CBS Cable simply a very bad business decision? Why did CBS insist on entering a business (televised culture) that the advertising community — a key link in the plan — nearly unanimously condemned?

The failure of a broadcasting giant's entry in the cable field sent a momentary shudder of dread throughout the cable industry. If CBS couldn't succeed in cable, could anyone? That question lingered in everyone's mind in fall 1982. Some concluded the failure demonstrated that culture on cable was dead. To examine the validity of these opinions, the development of CBS Cable needs to be traced from its beginnings.

On the evening of October 12, 1981, CBS Cable was launched with the appropriate pomp. A lavish party was held at the New York Public Library. The chairman was pleased and proud. In an interview published shortly after his retirement, Paley pronounced CBS Cable "one of the best things CBS ever produced. It was done beautifully, with good taste," he added. The pride in CBS Cable began with the service's debut, and in the days that followed, critical acclaim rolled in. CBS Cable was an instant and, as it turned out, an enduring critical success.

The Programming

The service had been announced in May 1980, and Jack Willis arrived as programming vice president the following November. "Four weeks after I arrived, I was told, 'Mr. Paley wants to see your fall schedule,' " recalled Willis. "I said, 'I can't do that, but I can tell him, conceptually, what I think we ought to be doing.' I made a presentation to Mr. Paley in which I told him what I thought we ought to do about identity, and then I gave him an example of a week's programming. I sold him on names: Makarova in *Swan Lake*, Jane Alexander doing *Calamity Jane*, Bernstein's Beethoven, and Twyla Tharp. I sold everything on names; that's how this business works.

"At a party later that night, Gene Jankowski (president of CBS Broadcast group with overall responsibility for CBS

Cable at the time) led me to believe that the meeting was the programming 'go or no-go decision.' "

Reflecting on his experience with CBS Cable, Willis said, "This is the first time I have ever been involved in something that turned out exactly as we planned it. We wanted to establish a strong identity for the channel, using a host, set, and graphics. In content, there was to be a balance — the high culture created by Merrill Brockway and more pop-ish material done by Gregory Jackson. We mixed relatively low-cost and small original productions with lavish, big acquisitions like *Swan Lake* and *Hamlet*, so the overall look would be big and fresh, even though most of the in-house production was small in scale. I have been involved with other ventures that were critically acclaimed, but I think this is the best thing I've ever been associated with, and I think it probably had much to do with the people I was working with, particularly Merrill Brockway, Gregory Jackson, and Regina Dantas, who was in charge of acquisitions."

The programmers wanted the service to have a look so clearly identifiable that it would be recognized by someone spinning the dial. At one point, according to Brockway, they even considered having a small red dot on the screen. Although the dot idea was abandoned quickly, the overall consistency of "look" was pursued throughout, and to a great extent, it was achieved.

Each evening was planned so that one program flowed nearly seamlessly into the next. Patrick Watson, the ubiquitous, tuxedoed host, introduced each program segment, sometimes in humorously irreverent style, and returned at the end of the evening to sum up and comment.

The interview program, "Signature," also lent continuity. It was a stunning talk show, remarkable because the host/interviewer, Gregory Jackson, was never seen. Only the guest appeared on camera, mostly in extreme close-up shots which created a level of intensity and involvement seldom offered to television viewers. Indeed, "Signature" may emerge as the most well-remembered aspect of CBS Cable's entire programming effort.

The major artistic triumph of the CBS Cable programming, however, was the work done by Merrill Brockway and staff to bring dance and theatre to television. Brockway had created "Dance in America" for PBS. As Willis pointed out, Brockway "had already been working with both Ballanchine and Tharp, and consequently, they were able to take dance even farther than they had done on 'Dance in America.' " Willis and Brockway were justifiably proud of the dance programming, which included works by Ballanchine, Tharp,

CBS Cable was "one of the best things CBS ever produced. It was done beautifully, with good taste."
—William Paley

...even the most hard-bitten television critics were forced to admit that CBS had broken new ground.

and May O'Donnell, along with imports and acquisitions that included work featuring Merce Cunningham, *Romeo and Juliet* with Galina Ulanova and the Bolshoi Ballet, the Royal Ballet's *Swan Lake* with Natalia Makarova and Anthony Dowell, and *La Fille Mal Gardée* performed by the Royal Ballet with Lesley Collier and Michael Coleman.

Drama on CBS Cable mixed first-rate imports with material produced by CBS Cable itself. Originally produced drama included Elizabeth Swados' *Songs of Innocence and Experience* and Athol Fugard's Tony Award-winning *Sizwe Banzie is Dead*, starring John Kani and Winston Ntshona, who had originally created the roles. Pat Carroll's one-woman show, *Gertrude Stein Gertrude Stein Gertrude Stein*, written by Marty Martin, and Robert Patrick's *Kennedy's Children* were remounted for television as was *The Resurrection of Lady Lester*, a work described by its author, OyamO, as a "dramatic mood-song," about the jazz saxophonist Lester Young. CBS Cable co-produced a stunning television version of David Storey's *Early Days*, starring Ralph Richardson and the original National Theatre cast from London.

Imported drama was a standout on CBS Cable, too. John Osborne's *A Gift of Friendship* starred Alec Guinness and Michael Gough. Ibsen was represented by *Ghosts* and *Hedda Gabbler*, which was adapted for television by John Osborne and starred Diana Rigg. Trevor Nunn's television version of Chekhov's *The Three Sisters*, with Royal Shakespeare Company actors, and RSC's *Macbeth*, with Ian McKellen and Judi Dench, were shown. CBS Cable was truly a feast for the drama lover.

Music was less successful, although the Bernstein/Beethoven series was outstanding. In a 1982 summer interview, Jack Willis conceded that "we are trying to figure out how to present music on television. I don't think we've even come close yet." At the time, Brockway was working with a number of jazz consultants, trying, as Willis put it, "to crack the format." Viewers never had a chance to experience the results, as the service closed down shortly thereafter.

CBS Cable acquired *Stravinsky*, a three-part musical biography that was an RM Production/London Weekend Television co-production, and *Piano Players Rarely Ever Play Together*, a jazz piano documentary by Stevenson J. Palfi which featured three generations of New Orleans jazz pianists. Demonstrating a sense of humor about music, The Kraft Music Hall presentations included an acquired production of *Carmen*, starring Grace Bumbry and John Vickers with Herbert Von Karajan and the Vienna Philharmonic, and CBS Cable's own production of *The Ring of the Fettuccines*, a

humorous spoof of grand opera featuring the Baroque Opera
Company.

Praise for CBS Cable's programming was nearly universal.
Although some critics were displeased by the lack of adventure in the transfer of art forms to television, even the most
hard-bitten television critics were forced to admit that CBS
had broken new ground. Television critic Brian Winston, for
example, admitted, "I think that at least Jack Willis
understands the conceptual dimensions of the problem of the
transfer."

Others called the programming too staid. They said it lacked
representation from the avant garde. That was certainly true,
but the programmers knew what they were doing. They were
establishing their expensive, classy beachhead. "We pitched
our programming to establish our identity and to co-opt the
field, to establish ourselves as number one with the audience,
the advertiser, and the artistic community," explained Willis.

"The first year concentrated primarily on performance aspects of culture, but plans had been made for the second
year's programming to be broader." Bill Moyers' "Walk

Twyla Tharp's *Baker's Dozen* from
the CBS Cable program "Twyla
Tharp and Dancers." (Photo:
Steven Caras.)

"The cable network was born, in fact, to counter a public relations blunder at the corporate level."
— Les Brown

through the Twentieth Century" was one example.

"The problem," according to Willis, "was how to justify programming that would bring in a broader audience within a cultural, performance network. I rationalized it by saying that if a program was good, all I needed to do was find a way to package it. This first came up when I saw the documentary, *American Challenge*. I decided we'd have a series called 'One of a Kind,' and we'll put anything good enough into it, and if the audience trusts us, and Patrick says it's worth watching, people will watch. In fact, *American Challenge* ran the first week, and Patrick introduced it saying something like, 'You wouldn't expect to see something like this on a network like ours, but it is one of the most thrilling films in the art of adventure, and we dare you to turn it off.' We did that on a number of programs, and it worked. And we planned to go broader on that basis. Our theory was that as long as a work was good, we could use it."

How It All Started

The idea for CBS Cable reportedly originated with Charlotte Schiff-Jones. She had taken the idea for a cultural service to Time Inc. before approaching CBS. When she went to Chairman Paley, she said to him, "You *must* be talking about cable; I have some ideas I'd like to discuss with you." Following that meeting, Schiff-Jones was asked to consult for CBS on how the corporation should enter the cable field. At a February 1980 planning meeting held in Rye, New York, she presented six programming alternatives, one of them for a cultural service. The corporate group guiding CBS's entry into cable explored a number of options. Some were more low budget, but the cultural option prevailed.

There were several factors that made an entry by CBS Inc. into cable seem attractive. It was 1980, the first year of cable's gold rush, and Wall Street was pressuring the big communications companies to get into cable. Everyone was being urged onto the bandwagon, and those who didn't play along were branded dinosaurs.

Les Brown, in his article "Who Killed CBS Cable" in the November/December 1982 issue of *Channels*, said, "The cable network was born, in fact, to counter a public relations blunder at the corporate level." When CBS fired John D. Backe without any explanation, the press concluded it was because Backe, who had been talking up cable, had clashed with Paley, who was committed to broadcast. "This made for a tidy story, but the press's assumption was dead wrong," said Brown. "Paley and the CBS board had *wanted* Backe to spread the word about the company's interest in cable, because security analysts" had been accusing CBS of being behind the times in new technologies. "The press coverage of the Backe dismissal was a public relations debacle for CBS, since it conveyed the wrong message to the investment community. . . . In a matter of days, CBS Cable was born." CBS already had a task force dealing with cable questions. It was easy to let the task force go ahead.

The whole idea even made sense: The major U.S. cities were just being franchised, and visions of a new, upscale, sophisticated cable audience danced in the heads of planners. CBS Cable, a cultural service, would be the perfect offering in the new urban cable markets. It was almost a necessity as soon as you thought of it. Added to this was the fact that everyone knew that Paley wanted a quality item like this.

Furthermore, CBS officials felt that they could tap into the potentially valuable PBS audience. "Imagine how much you could make if you could advertise on PBS," suggested one early planner. "If you assume there are 30 million viewers and PBS gets an average rating of 3.5, if CBS could do that, it would be good business."

CBS did not conduct a feasibility study before entering the cable business. Rather, Charlotte Schiff-Jones described the assumption underlying CBS Cable as a "belly-button" feeling.

Thomas F. Leahy, who as executive vice president of CBS Broadcast Group had corporate responsibility for CBS Cable, was asked in an interview for this book why CBS Inc. had not conducted a feasibility study before entering the cable programming business. He replied, "CBS Cable was a small business by design, so you don't *have* to get broad reaction. The concept was tough to test because it was brand new, and the situation with cable operators was evolving all the time, on the hardware front, and on the satellite front. Everything was constantly changing, and as one piece changed, the others changed in their relationships. It was not a situation that we could isolate, put under a microscope, and look at."

When pressed with the information that ABC had spent a million dollars and over a year on research prior to its entry into the cable field, Leahy conceded, "Well, we had our research, and our research was in terms of audience reaction, and we were committed to making a major statement. Clearly, we were ahead of our time."

CBS's strategy, Leahy explained, was "to be the first company to identify itself in the public eye as the quality service, thereby establishing a beachhead. Because the audience universe was so small, it would be too expensive for another company to come in with a challenge — to fight for a portion of

Charlotte Schiff-Jones described the assumption underlying CBS Cable as a "belly-button feeling."

an already small audience."

CBS may have thought it had developed a good, if small, boutique-like business to run along side its sprawling broadcast empire. Leahy insists that was the case. The concept had been received with genuine enthusiasm by oil companies and others. And, CBS apparently did think that it needed to establish a beachhead, although it was beaten to the punch by two other cultural networks.

The most convincing reason for CBS to go into cable, however, appears to be a negative one. The company didn't want to be branded a dinosaur.

Although there was no feasibility study, a business plan with estimated revenues, expenses, and break-even projections had to be submitted to Chairman Paley for his approval before the service could receive a go-ahead. "They made their first mistake," commented a former CBS Cable executive, "when they estimated the advertising revenues for that business plan. They approached cable ad revenues the same way they would have for broadcast advertising in their revenue projections." The numbers that resulted from that model were acceptable enough to get the CBS Cable business plan past the notorious Paley eye for the bottom line. But those figures eventually played a major part in the CBS Cable story.

Planning completed, the service was launched to extraordinary applause. But in the words of CBS Cable's former president, Richard Cox, "We had seven weeks—three in October and four in November—to enjoy our critical success. From then on, it was all downhill. It was just one damn thing after another."

The Pay-Basic Debate

A major decision made during the early planning phase was that the service would be basic, not pay. CBS Cable had been proposed as a pay service originally, an idea which Paley, himself, is said to have rejected, arguing "You can't chase HBO and succeed."

The pay or basic issue was examined before the service was announced, and when the early decision in favor of basic was made, "it was unanimous," reported Aaron German, CBS Cable's former vice president for finance. There were good reasons for launching as a basic service. Viewers needed time to find the service and sample it without having to pay extra. And CBS needed time to develop a market for the product, which would be unknown at the time of its debut.

The summer before the service was launched, however, the

pay-basic issue was raised again, and from that point forward until its shut-down was announced, the decision was never quite firm. The option of going pay was held open until the end. The pay-basic debate is one of the themes that weaves in and out of the CBS Cable story—appearing, disappearing, and then reappearing. Staff opinions on the topic were strongly felt, and at times it seemed as though open warfare had broken out among the CBS Cable staff.

From late fall 1981 on, it was clear that CBS Cable was in trouble as a business. More revenue was needed, and interest in converting the service to pay intensified at this point. Chairman Paley was reportedly almost certain by November that the service should go pay. But the figures didn't add up right. Paley kept sending the cable team back to re-examine its conclusions, but some people think he never got the figures he seemed to want. The CBS Cable financial projections never gave an unequivocal and indisputable indication that going pay would succeed or fail.

Although going pay would have opened up a new revenue stream, it would also have caused complications and increased expenses. A chief consideration was the cable operators who had already signed contracts to carry CBS Cable. A switch from basic to pay would mean renegotiation of all those contracts and a possible loss of many of those affiliates. Certainly some of the cable companies that welcomed CBS Cable as a free service would have no interest in it as a pay service.

Reintroducing CBS Cable as a pay service would not be a straightforward matter, even for many cable operators who might want to do so. Few systems carrying CBS Cable had the kind of addressable equipment that makes introduction of new pay services easy. Unlike basic services, which go to all subscribers, pay service signals must go only to the subscribers who pay for them. Some operators do this by sending out a scrambled signal which is decoded by a converter, while others install a device called a trap to screen out the signal. Both methods require a trip to the subscriber's home, a labor intensive and expensive service for the operator to perform. In interactive, addressable systems, the headend computer easily does all the work.

Changing from basic to pay status would have increased CBS's expenses as well. Some felt more programming would be necessary to support a pay service. Others disagreed, contending that CBS could change to pay in exactly the same form as it had premiered, even retaining the same repeat schedule.

Everyone agreed, however, that the actual *production* of programming would become more expensive. Union scales,

The pay-basic debate is one of the themes that weaves in and out of the CBS Cable story. . . .

Athol Fugard's play *Sizwe Banzi Is Dead*, with actors John Kani and Winston Ntshona, offered a wry and sardonic comment of life under apartheid and the passbook system to CBS Cable viewers. (Photo: Leonard Kamsler.)

where established, were higher for pay cable services than they were for basic services. While many basic services had opened without union contracts governing some areas, a number of unions had negotiated contracts with HBO that had become models for pay cable.

Thus, going pay would have been complicated for CBS Cable. But it would have added revenue. Or would it?

There was genuine disagreement over whether going pay would actually succeed at all, and that disagreement became rancorous at times. Some participants in the discussions charge that the figures used in presentations to Paley and Wyman on the basic versus pay question had been "manipulated" to give the worst possible case for going pay and the best case for staying basic. Another said that "even with rather optimistic projections on the number of consumers who would sign up, and the amount of systems that would accept it as pay," conversion was not a "workable" option; "it would not pay out."

One industry observer, an expert in the operation of pay services and familiar with CBS's pay plans, commented that CBS Cable would not have succeeded as a pay service. "The programming costs were far beyond what you could expect in revenue." And, the observer added that their penetration estimates were way too high for a service with such a narrow appeal.

The pay-basic debate was never fully resolved and continued throughout the entire span of CBS Cable's existence. A continual cause of friction among executives and source of doubts in the advertising and financial communities, the debate was played out in the trade press, adding to the problems of an already harassed and troubled ad sales force.

Advertising: An Attempt to Return to the Fifties

CBS Cable's ad sales people faced a number of frustrations while trying to make the service truly ad-supported. The gap between estimated and actual sales revenues is an index of the impossibility of the task. Early business plans for CBS Cable called for approximately $40 million in ad sales. The actual revenue was closer to $8 million.

The advertising plan called for selling program sponsorships, an approach different from current Madison Avenue practices. The idea harked back to the fifties, when program sponsorships were a common form of advertising. But times have changed. Today, advertisers buy thirty- and sixty-second spots. They don't buy ad "environments." They don't buy sponsorships either. And to make matters worse, the people who were familiar with the fifties approach to advertising were long gone from the ad agencies by the time CBS Cable arrived.

Thomas Leahy explained the hopes of the CBS Cable planners this way: "Here was the opportunity to do 'Television 1958,' when Stoppette had their sign in front of John Charles Daley for half an hour straight." Back then, Leahy explained, people paid for certain environments, and CBS wanted to return to that approach.

"CBS Cable was intended to be a complete marketing tool,"

Early business plans for CBS Cable called for approximately $40 million in ad sales.

Leahy added, "not simply an advertising vehicle at cost-per-thousand. A sponsor would buy an entire show and use the sponsorship very much like Mobil markets its sponsorship of Masterpiece theatre. Mobil spends as much money telling people about what they are doing as they do showing the viewers the product. It turned out that was not the way the advertising community wanted to use CBS Cable.

"The oil companies, after showing considerable initial interest, dried up. General Motors and the other automobile manufacturers were having serious problems at that point, too. So the logical big-ticket items — the people who had some history with this kind of marketing strategy — were out of the marketplace."

Some have observed that since ad agency practices had changed since the fifties, the appropriate sales targets for CBS Cable sponsorships were not agency media planners, but chief executive officers. "It was not necessarily the advertisers who didn't come through; it was the *ad agencies* that refused to buy environments," commented one observer.

The early versions of CBS's ad plan also included provision for advocacy advertising, which may have increased the service's attractiveness to oil companies during early discussions. Later, CBS decided not to allow advocacy advertising, and although some at CBS downplay the importance of that decision, others contend that it caused significant harm to the channel's ability to sell. "The decision that there would be no advocacy advertising certainly didn't help," concluded one ex-official of the service. It surely made CBS Cable a less attractive buy for the oil companies and others who wanted to reach an affluent, influential audience.

The ad sales effort started in May 1981, nearly six months before the service's debut. By December, it was clear that the strategy was not working. Robert Turner, the former general manager, explained: "Our sales problems turned out to be that the unit costs were *much* too high. The assumption was that the market would pay five times what they would ordinarily pay. As it turned out, they would pay one or two times the ordinary price. Only a very small number would pay five times as much, and those who would — like Kraft — stepped forward. That continued the illusion that our goal was achievable. By the time all the rest started sorting itself out, a lot of valuable time had been lost."

"We lost some revenue and much of our credibility with Black Rock [the nickname for CBS Inc.'s corporate headquarters] while we debated strategies internally and tried to maintain the sponsorship approach with advertisers," explained one former official. Another member of the CBS

Cable team described his frustration, "We kept re-doing our budgets. We could hardly deliver *anything* in sales. It got to the point where it was a weekly experience for us to go up to Black Rock and say 'Remember that number we gave you last week? Well, forget it.' " All in all, the delay in changing ad sales strategy cost CBS Cable dearly. Credibility was lost in the upper reaches of corporate management and on the outside with advertisers.

James A. Joyella, vice president for advertising sales picks up the story: "There was a determination that we would get a premium for our product, and that had the effect of pricing us beyond what the marketplace wanted to spend. We wanted to be sure that we had run with that plan as hard as we could before we walked away from it. So before we adjusted our prices downward, a lot of agencies made presentations on CBS Cable to their clients. Then, when our prices came down, many advertisers had already been convinced that CBS Cable was too expensive for them. As a result, we had to go back and re-sell them. It was frustrating as hell."

CBS Cable's skidding prices made selling difficult, and other factors seriously affected the sales staff's credibility with advertisers. Richard Cox, CBS Cable's former president, explained, "Rumors that CBS Cable was going pay were the first obstacles to selling. 'We hear you are going to go pay,' they would say. 'No we are not going to go pay,' our people would reply. The first half hour of every sales call became how we weren't going pay."

The next thing was the proposed joint venture between CBS and Twentieth Century-Fox, which was originally supposed to include CBS Cable. "They would say, 'Fox isn't going to go along with CBS Cable the way it is now. What kind of a service are you going to be with Fox?' 'We don't know if there is going to *be* a Fox deal,' we would reply.

"And then, after the Fox deal fell through, for the next six months they wanted to know, 'Well, are you still going to be around?' "

Worst of all was the fact that from December on, "I couldn't get anybody above me to come out and say, 'Yes, CBS Cable is going to stay around,' " said Cox. "Nobody wanted to get caught saying that." Rumors abounded, and there was virtually no support from the top level at CBS Inc. All this added up to nearly insurmountable odds for a sales staff.

Once the changeover was made from sponsorships to selling spot time, there were additional difficulties. "If you sell sponsorships," explained Cox, "you can price them an entirely different way than if you have to measure—bean for bean —against WNEW, CBS television network, and all the rest.

"They priced the ads as though they were going to the chairman of the board, and then they tried to sell them to people who don't think that way."
— Charlotte Schiff-Jones

So once we got into the selling of minutes, which clearly we had to do, we had to adjust our prices, and we weren't competitive dollar-for-dollar. Further, as a result of those changes, even if we sold as much as we intended to sell, we would have been getting less money for it."

Some observers still feel that the sponsorship approach to ad sales was the proper one for CBS Cable to follow, and that the foul-up came not in the strategy, but in the implementation of it. Said Charlotte Schiff-Jones, "They priced the ads as though they were going to the chairman of the board, and then they tried to sell them to people who don't think that way. When the strategy changed, the ad sales people went to a totally conventional approach with none of the conventional ammunition. That approach was doomed."

Selling was certainly tough under those conditions. Advertisers wanted numbers, but CBS needed time to establish those numbers. With a sales argument that contended, as CBS Cable did, that for the first time, the advertiser had an opportunity to reach the light TV watcher or the person who doesn't ordinarily watch television, time was required to demonstrate that those who don't usually watch television are now watching it. It was an interesting argument — a chance to catch people who have not watched television in the past, who could only be reached through *The New Yorker*. But how could the company demonstrate conclusively that non-watchers had been turned into watchers? It required time. And that is what CBS didn't have for the CBS Cable ad sales staff. Time too do the research, and time to let the public know about the service. All of that could not be accomplished in eleven months.

"The question, 'What is the market for cultural cable?' was never answered at the time the planning was done," contended a member of the sales division. "You cannot create a market for something just because you *want* to do it. You can only exploit a given appetite. We were assuming there was a need and went out to fill it. But as we got further into selling, we found the need was not so great. Our proposition was interesting and appealing, but there was not so great a need as we had thought originally."

Marketing CBS Cable

The marketing staff's mission was to convince cable operators to carry CBS Cable. The marketers faced two major hurdles: CBS was on the "wrong" satellite, and ARTS, a competing basic cultural service, had been launched a full six months ahead of CBS Cable, thereby getting a jump in the race for scarce channel space. In spite of these obstacles, CBS

Cable was ahead of its subscriber goals when the service closed
down. When it began, the service had a potential viewership
of 1.5 million, and when its demise was announced, it had 5.5
million, nearly 1 million ahead of the goal set for the service.
The growth rate had been steady and impressive, despite
rumors that the service would go pay or close down.

CBS Cable was transmitted from Westar IV, not Satcom
III, the satellite received by most cable systems' earth stations. Consequently, most cable system operators could not
receive CBS Cable without purchasing a new earth station.
To help rectify the problem, CBS Cable gave earth stations to
some cable operators, but not to all who signed on for the new
service. "Earth stations were given only in situations where
there was a major number of subscribers to be gained," explained former marketing vice president Schiff-Jones.

CBS also enhanced its own attractiveness by assisting
operators in marketing the service to subscribers. This was
done in several ways. Operators received up to fifteen cents
per subscriber to subsidize their efforts in marketing the service. CBS was the only cultural service to provide this kind of
assistance to cable operators.

CBS Cable was also the only cultural service to advertise in
the consumer press and to undertake creative marketing projects designed to gain exposure and support for the service in
the artistic community. "All our consumer advertising was
barter," explained Schiff-Jones. "We didn't pay a dime for it.
People kept saying, 'My God, you're advertising in *Newsweek*,
in *Saturday Review*. You're spending money like water!' But we
had millions of dollars in advertising bartered for spots on the
channel. We were ideal for the *Saturday Review*, because our
audiences were the same. *Newsweek* was making special commercials for CBS Cable exposure. Unfortunately, *Newsweek*'s
were never seen; they were holding off until our universe was
larger, but we were gone before they were ready."

CBS marketers also scored a major triumph when they
won acceptance for the service in Tele-Communications,
Inc.'s (TCI) national program package. In 1982, TCI, with
approximately two million subscribers the nation's largest
MSO, announced creation of a national package of program
services to be offered by all its systems. Because inclusion
would instantly add two million subscribers, all program services were vying to be included.

TCI and CBS Cable officials had reached an agreement
that would have brought CBS the additional two million
subscribers at a crucial time. But the deal fell through when
TCI demanded assurances that CBS Inc., the service's corporate parent, would not make.

CBS Inc.'s top management refused to even make a phone call saying the service would survive.

John Malone, president of TCI, explained his reasons for the company's demands: "I was concerned about CBS Cable's viability. I said I would put CBS into the national package if we could get a legal commitment from CBS, with some real teeth in it, that the service would still be in existence for a few years into the future. I didn't want to make a false start, and I was almost paranoid with concern that they were uneconomic.

"So we said to them, 'You bet, you're in our national program package, and the only thing you have to do is give us some legal assurance that you are actually going to do what you say you are going to do.' I insisted that CBS had to satisfy our attorneys that the corporation would take some kind of a substantial hit if it didn't fulfill its verbal commitments. As I told my staff, 'Money talks and bullshit walks, so have CBS put it on paper.' "

CBS Inc.'s top management refused to even make a phone call saying the service would survive. When asked why, Thomas Leahy, CBS Broadcast Group's executive vice president, explained, "We couldn't let John Malone feel he had our commitment, because we weren't one hundred percent behind the service, and we had too much respect for our relationship with him. I knew that the evaluation process was in place, and I was falling off the complete support level at that point, and I didn't want to lead John on."

The TCI request for assurances came at the end of April 1982. "That was a painful moment of truth when corporate refused to make the phone call," remembered Schiff-Jones. "That was when Bob Turner and I surmised exactly what was happening."

CBS Cable had become enormously popular. By determined efforts, a talented marketing staff had largely overcome the handicap of being on the wrong bird. But, at the very moment that CBS Cable executives were scoring a victory that other services would have nearly killed for, executives at Black Rock were deciding to pull the plug on the six-month-old service.

Joint Ventures

If it is true (as most former CBS Cable executives agree it is) that by December 1981 it was clear that CBS Cable could not live up to its original business plan, and that the prospects for significant advertising revenue were bleak at best, then joint venturing — sharing the financial losses with another company — might have provided a solution. CBS Cable explored a series of joint venture possibilities during its life, none of which came to fruition.

The first of these, according to Richard Cox, the service's former president, took place in November 1981. Almost immediately after CBS Cable was launched, Charles F. Dolan, part owner of Bravo, contacted Thomas Leahy about merging Bravo and CBS Cable. Although nothing came of the early talks, the discussions were revived as CBS was about to go under.

As 1981 drew to a close, CBS Inc. began talks with Twentieth Century-Fox Film Corporation about a joint venture in the home video and cable business. CBS Inc. apparently thought Fox was a good prospect to share the burden of CBS Cable. The talks went on for nearly six months. During that time the doubts about CBS Cable's viability as a business mounted.

Also during this period, Stephen Roberts, who had been named president of the proposed joint venture, had taken the reigns at CBS Cable, and all financial decisions had to be approved by him. As far as the cable executives could gather, the joint venture was a *fait accompli*. Eventually, CBS and Fox reached agreement on the terms of their venture. But CBS Cable was left out. Fox was unwilling to pick up any of CBS Cable's mounting losses.

In retrospect, it is clear how important the Fox partnership was to CBS Cable's overall prospects for survival. CBS President Thomas Wyman had made it clear at CBS's annual meeting in April 1982, when he said that the joint venture was central to the CBS development strategy.

According to Cox, it had become increasingly clear during the negotiations that, had CBS Cable been included in the deal, it would have been changed into a service that more closely resembled WTBS, the superstation, than its original format. "Our discussions were all in the direction of, 'How can you get stuff in here that will make money?' " he said. Executives were making jokes about becoming the "Gomer Pyle Network."

It is not difficult to understand why CBS would want a partner to help absorb cable's initial losses. What *is* difficult to understand is the choice of Fox. Fox was admittedly cash poor, and a look at the final terms of the agreement between the two companies highlights the apparent futility of hoping Fox would sink itself deeper in debt. The final agreement between Fox and CBS (as reported in the *Los Angeles Times*, July 14, 1982) called for CBS to provide $45.5 million in interest-free, twenty-year loans over the next two years. Fox is reported to have said that it would receive $16.75 million on signing the deal, $16.75 million after a year, and $12 million after the second year.

Executives were making jokes about becoming the "Gomer Pyle Network."

Of course, CBS Cable, a cultural money loser with narrow appeal, was dropped from the deal. Given Fox's financial situation, how could it have been otherwise? Fox didn't want anything to do with an additional revenue drain. The question is: How could CBS have ever thought it was possible?

The final agreement between Fox and CBS established a joint venture which was to combine the video cassette operations of both companies. The CBS Studios at Studio City, California became the property of the joint venture. There had been rumors during the early CBS-Fox discussions that what was really behind the deal was Fox owner Marvin Davis' desire to turn the Beverly Hills Fox lot into a real estate development. In order to do that, of course, Fox would have had to have other studio facilities.

If there was any truth to those rumors, the CBS studio property would have been exceedingly valuable, perhaps even essential, to the success of Davis' plan. Early in 1982, when the magnitude of CBS Cable's ad revenue shortfall had not yet become clear, it might have been reasonable to argue that Fox accept CBS Cable in a deal that would liberate the Beverly Hills property for lucrative real estate development. As the magnitude of the CBS Cable losses began to emerge, Fox's interest cooled to the point where Stephen Roberts was quoted in *The New York Times* as saying he was " 'not sure' that CBS Cable in its current form can make money. 'I don't know too many people who would want to invest in a business if it can't make money,' " he reportedly added. Furthermore, few real estate deals remain attractive when interest rates soar above 20 percent, as they did during that period.

It appears as though Fox was unwilling to consider including CBS Cable in the joint venture unless the character of the service was changed, and CBS Cable executives certainly thought some programming changes were in the works at the time. Leahy said that there may have been conversations about changes in the service, but "no *decision* of that nature was ever made." Although CBS Inc. may have been willing to hold discussions with Fox about changing the service, that same management steadfastly refused to consider altering the character of the CBS Cable service after the Fox deal left cable out in the cold.

Once it was decided that CBS Cable would not be part of the Fox deal, management at Black Rock went back to the drawing board on going pay. Apparently, top management had decided that the wait for a financial turn-around was going to take too long. They determined that the only way the service could possibly work as they wanted it to was if another company would take over the sales and marketing of CBS

Cable as a pay service on a partnership basis.

   Aaron German recalled, "In one of our last presentations to Messers Paley and Wyman, our analyses showed that basic wasn't going to work (we had known that for a long time). Pay, going it alone, wasn't going to work unless we had a partner who was an MSO, who would, in effect, guarantee distribution. The only way we felt that we had a shot at it was if we could convince an HBO or a Spotlight, for example, to join us."

   HBO had all the necessary qualifications. Discussions were held with HBO during June and July 1982 in an attempt to interest HBO, or its owner Time Inc., in marketing CBS Cable as a pay service. The talks between CBS Cable and HBO did not go very far, even though some at CBS Cable had cherished real hopes that such a venture might save the doomed service. Although Time Inc. was clearly interested in CBS Inc. (witness the eventual deal announced in December 1982 that joins CBS, HBO, and Columbia Pictures in a plan to create a motion picture company), there was little enthusiasm at HBO for CBS Cable as a business. CBS never created a deal attractive enough to convince Time Inc. to accept CBS Cable as a "necessary evil" to make possible a more desirable venture between the two.

   CBS Cable's fate appeared to be sealed, from Black Rock's

Short Stories, choreographed by Twyla Tharp to music by Bruce Springsteen, from *Confessions of a Cornermaker* on CBS Cable. (Photo: Steven Caras.)

CBS never created a deal attractive enough to convince Time Inc. to accept CBS Cable as a "necessary evil."

point of view, from the time that the Fox deal fell through. The possibility of a joint venture with Time Inc. briefly revived hopes, but when HBO eventually said no, the CBS corporate parents appear to have let go of their child permanently.

At CBS Cable, however, staff members worked feverishly to come up with another joint venture possibility that might save the service. Marketing vice president Schiff-Jones contacted Times Mirror Cable TV, Inc. about joint venture possibilities, and Times Mirror referred CBS Cable on to the management of Spotlight, a pay movie service of which Times Mirror is part owner. The concept of a joint venture with Spotlight at first looked promising. According to finance vice president German, "What made Spotlight potentially exciting was our information that the service was about to be launched outside of its MSO ownership's cable systems. Our thinking was: Spotlight can market us at the same time. We'd be a natural addition, and it would add very little incremental cost to them and, consequently, to us." Nothing came of the talks, however, because unknown to CBS Cable, Spotlight itself was mired in financial problems.

Serious talks took place with Bravo as well. The November 1981 talks about a merger were revived during summer 1982. "The Bravo deal was very close to taking place," said German. He and Lawrence Hilford (subsequently named president of CBS-Fox Video), who was negotiating for Bravo, could not come to an agreement on whether the joint service would be marketed by CBS or by Bravo. "We just couldn't resolve the issue within the time we had," said German.

Informal talks were also held with Showtime, and at the end of the road, a move was afoot to charge operators carrying CBS Cable a ten to fifteen cent per subscriber charge. "There was a corporate meeting out at Glen Cove in September 1982," recalled Schiff-Jones, "and I kept telephoning them, letting them know we got another million dollars, as the commitments came in from the cable operators. But they barely tolerated my interruptions. It was really a desperate last-ditch attempt—a kind of death rattle. But the support voiced by the operators was incredible. We had over $5 million promised by telephone."

But the decision had been made by then. CBS Cable was doomed. The rest was just a formality. And the formalities were apparently being observed at those same Glen Cove meetings. Several weeks later, the staff and the public were informed that CBS Cable would close down.

Why Did CBS Get Out of the Business?

CBS Cable was abandoned for a number of reasons, but

certainly not because of the size of the first-year losses. They amounted to no more than the network would have lost from the cancellation of two prime time series. However, 1982 was a difficult year financially for nearly all CBS's divisions. Wall Street was troubled by CBS Cable's mounting losses, and William Paley was going to retire, at long last passing control of the company to a new man.

At the time CBS Cable was closed down, the only reason cited in the official CBS Inc. statement was that "advertising revenues were well below expectations." Thomas Leahy, in June 1983, said that the company pulled the plug because the losses were going to continue for too long. "A deliberately small business can only afford a certain number of years in deficit. For CBS Cable, it was determined originally that the number of years was four. When it began to be clear that it would be far more than that, it was concluded that to continue would have been an irresponsible investment," he explained. When pressed to reveal how many years the revised CBS Inc. projections showed it would take for CBS Cable to break even, Leahy declined to give a specific number.

A number of observers have commented that the corporate decision to close down CBS Cable is not difficult to understand, because "a credible business plan was never presented to the highest levels of management at CBS Inc." Many agree that weak leadership at CBS Cable, along with the division's inability to recover from poor initial planning, contributed to the service's failure. One former executive from CBS Cable pointed out that "the internal fighting never allowed us to agree early enough on a workable business plan. That kind of confusion made it easier for corporate to kill us. We no longer had any credibility."

CBS Inc. had had a terrible year. The theatrical film unit had lost $15 million. The records division was in a dramatic decline. Three hundred were laid off from the division, and in a separate action, a plant was closed, eradicating 1,250 jobs. Other divisions were not so badly troubled, but there were no standouts in 1982.

The Wall Street analysts who had been so eager to see CBS involved with new technologies had become increasingly concerned over the cable service's losses. CBS Cable had become a well-known money loser in the business press. Mention of the service always carried a dual identification tag that read: "critically acclaimed, deficit ridden" or something similar. Wall Street pressure surely was a major contributor to the pull-out. In a significant move, CBS's stock rose 2 and 5/8 the day following the announcement of the decision to close CBS Cable.

> "A deliberately small business can only afford a certain number of years in deficit."
> — Thomas Leahy

Broadcast could hardly be expected to welcome the new cable baby to its bosom.

Furthermore, Thomas Wyman was about to assume control of the company. Paley had announced his intention to retire shortly before the final decision was made to close down CBS Cable, and Wyman's ascension to his post undoubtedly had something to do with the move. Wyman, who had started out as an enthusiastic supporter of CBS Cable, may have just wanted to get the agony over with, to kill it in its first year. Then the losses could be treated as an extraordinary expense in a year that had been tough all around for CBS. CBS Cable was a highly visible, easily isolatable problem. Removal was easy, and it would help Wyman start with a much cleaner slate. So it was done.

Why Did CBS Cable Fail?

The primary reasons seem to have been bad corporate planning and lack of support from the Broadcast Group, coupled with an inability on the part of upper management to respond quickly to changes in the cable environment.

Organizationally, CBS Cable was part of the Broadcast Group. It is not difficult to understand that Broadcast could hardly be expected to welcome the new cable baby to its bosom. Network television was poised on the brink of a cable-induced dispersion of audience, and there was a distinct lack of enthusiasm for cable among broadcast executives. Thus, the new operation started out with several strikes against it.

A number of damaging business mistakes were made by the people responsible for planning the new business. There was no thorough investigation of the market. The advertising strategy was idiosyncratic for the times, and the ad revenues were badly overestimated. Finally, CBS mounted and ran a high class boutique operation in an environment where street vendors were the dominant form of commerce.

The initial planning of CBS Cable was startlingly undisciplined for a company of CBS's reputation. The cable market was not well investigated. At the time CBS was planning its cable entry, there was a great deal of talk about wiring the cities. Despite the fact that it was about to invest hundreds of millions of dollars in cable, CBS seems to have taken cable's hype at face value. A careful examination of the prospects for speedy wiring of the cities would have revealed that it was not going to happen as fast as CBS hoped. That large, new, sophisticated urban audience was not just around the corner.

When CBS announced the closing of CBS Cable, the official company notice cited the fact that the "advertising revenues were well below expectations." Indeed they were. Early ad revenue estimates for 1982 ranged from over $40 million to approximately $20 million. Estimates continued to be revised downward after the service's launch. The actual revenue brought in over the service's eleven-month life was closer to $8 million.

How could CBS have made such an enormous error?

There are several explanations. Leahy says the oil companies, the ad agencies, and others encouraged them to go ahead. Schiff-Jones agreed. "They *really* did encourage us," she said. Apparently CBS expected, on the basis of their informal survey, that the ad support would be substantial.

Furthermore, impressive revenue projections were apparently necessary to convince the chairman to go into the business. Anthony Hoffman, a media analyst for A.G. Becker, said of the inflated ad revenue projections, "This was done solely to secure Paley's support. They [CBS executives] knew it was extremely important that the project go forward, and so they overestimated the returns. They had to get Paley to say yes."

Everyone knew Paley wanted to do quality programming, but he also had a ferocious eye for the bottom line. He had nixed unprofitable deals before, despite their appeal to his higher instincts. To get the business plan past Paley, explained one former CBS Cable official, "the figures *had* to be profitable. To talk CBS into investing that much money, you had to show how long it would take to make the money back, and it *couldn't be too long*." Indeed, CBS later said the cable service was shut down because it would take *too long* to reach profitability.

Yet another former CBS Cable executive theorized that the "mistake" may well have been intentional, but not malicious, a mistake made "to clear one corporate hurdle to get into the business." It may well have been done with the attitude, "we'll correct it later," he added. But correction later proved impossible.

It is puzzling that CBS Inc. did not consult with authorities in cable advertising. The warning signs perceived by others in the cable business — advertisers who scoffed at the idea of cultural cable and the ad community's slow acceptance of cable as an advertising medium — were apparently not perceptible at Black Rock. Or, perhaps the company simply had confidence that it would overcome such difficulties. After all, didn't CBS practically invent television?

Also worth noting is the fact that CBS Cable, as a business, was totally out of synch with the rest of the cable business. John Malone, president of TCI, commented: "Cable has always been a bootstrap business, and it still is to a degree, although its boots are getting pretty big now." CBS, being network television, had an "ability to spend five times as

The initial planning of CBS Cable was startlingly undisciplined for a company of CBS's reputation... CBS seems to have taken cable's hype at face value.

Dick Anthony Williams stars as jazz saxophonist Lester Young, and Mary Alice portrays singer Billie Holliday in CBS Cable's ninety-minute drama *The Resurrection of Lady Lester*. (Photo: Courtesy of Camera Communications.)

much money as it would take any other entity to do the same thing," he said. This is not just a cheap shot. Malone explains: "They were located in New York, the costs of their operation were high, [union costs for operations, for example, were admittedly high]. CBS has always worked in a program environment of much more expensive budgets and more elaborate facilities than cable was used to."

No one disputes that what CBS did, it did well, with class, flair, and quality. Experienced observers did question the level of expenses throughout the operation—not just programming, which when viewed in perspective was not *that* expensive—but the entire operation. The CBS Broadcast Center as an origination point, the Sixth Avenue offices, the seemingly enormous staff. Most cable operations that survive are run on a shoestring, cable executives like to point out. CBS had long ago forgotten about shoestrings.

Poor day-to-day strategy and a seeming inability to adapt to rapidly changing conditions seem also to have plagued the operation.

CBS Inc.'s initial inclination to include CBS Cable in a possible joint venture with Fox may have made good sense. But as the months elapsed and CBS Cable's losses mounted, why did Black Rock insist on pushing a money loser on cash-poor Fox?

Apparently there were many conversations about the way CBS Cable would have to be changed if Fox allowed the service into the deal. Some said it would not have remained a cultural service. Leahy said CBS Inc. would not have changed the character of CBS Cable, but others had the clear impression that change was imminent as part of the Fox deal. If Black Rock was willing to contemplate a "Gomer Pyle Network," as executives ruefully joked, why was management unwilling later to consider the CBS Cable general manager's plan to expand CBS Cable's service by adding desirable movies, travel, and gourmet food programming designed to appeal to the same audience? The most obvious reason for that refusal is the fear that affiliates would be too offended by a broadening of CBS Cable's appeal.

But if CBS Inc. was concerned about the attitudes of the affiliates, and a broadening of appeal was necessary to make CBS Cable a part of the Fox deal, were the Fox conversations just a sham? Or, would the fact that it was no longer just CBS, but now CBS-Fox, have mitigated the affiliate's ire and made the change acceptable? There are no definitive answers to these questions.

The preoccupation with Fox as a solution to the financial problems CBS Cable was causing management seems to have also prevented the company from looking carefully at other solutions. How can it be that the service was closed down as an agreement was being negotiated with Bravo? If those talks had been started sooner, could the sticking point have been overcome? Why were seemingly hopeful efforts to save or create other joint ventures put off until it was too late? It appears that CBS Inc. stuck blindly to the hope that Fox would solve the problems, even when that hope flew in the face of practical considerations. Or, were the later discussions with other companies (following Fox) exercises in futility, pre-

...why did Black Rock insist on pushing a money loser on cash-poor Fox?

Les Brown concluded that CBS Cable died from "lack of parental love."

ordained to failure because corporate had already made up its mind?

Les Brown concluded that CBS Cable died from "lack of parental love." The kind of difficulties that plagued the service's management from the time of launch are indicative of the lack of care devoted to the new venture. Determined words of support from corporate leaders at CBS would have gone a long way to silence the press's weekly gloom and doom treatment of CBS Cable. As it was, there was no visible support from above, and when mistakes were made at the CBS Cable level, as they were, they simply compounded themselves because of lack of leadership, clear policy, and most of all because of a lack of determined commitment from CBS Inc.

Does the Death of CBS Cable Mean that Culture Can't Survive on Cable?

The lessons to be learned by CBS Cable are lessons in management, not lessons about culture's ability to survive on cable. In fact, many in the cable industry who mourn the death of the service are quick to point out that a return in several years as a pay service makes good business sense. As John Malone said, "If CBS would come back into the industry in about a year on an addressability basis, be much more selective, target their audience, I think they'd be extremely successful."

The business was badly planned. Revenue estimates were so far off that it is very likely that they were intentionally misstated. Because operating budgets were developed on the basis of anticipated revenue, and the substantial start-up costs of the new business were incurred long before revenue began to appear from ad sales, it was impossible for revenues to catch up to expenses within the time allotted in the plans. Although expenses were cut drastically, CBS Cable started its life hopelessly behind in the race to break even. And, it appears that the aborted Fox deal was Black Rock's only serious attempt to solve that problem. Furthermore, if corporate support was being withdrawn after only six months (in April), why did it take another five months before the decision was final?

Given all these facts, and if Malone is correct, it is hard to believe that CBS Inc. was serious about succeeding with CBS Cable in the first place. That is not to say that CBS Cable was set up to fail, although that is possible. CBS Cable does appear, however, to have functioned at least partially as a Trojan horse to get CBS into cable, placating Wall Street without offending the affiliates. The service was more a response to outside pressure than a wholehearted try at a new business.

The Arts Transformed

When a concert, a play, or a dance is transferred to television, the work ceases to be a play, a concert, or a dance. It becomes television. Watching a televised production is fundamentally different from the experience of a live performance. "Television is not an art form and translation is not an art form, it is a craft," contends Kirk Browning, a highly respected film and television director who has been responsible for a number of "Live from the Met" opera telecasts. "We have to recognize that the marriage of the performing arts and television produces a 'monstrous Caliban' aesthetically." [1]

Certainly, the television monster is not to be feared for its size. Television routinely reduces the sweep of the Metropolitan Opera House to twenty-one inches or less. Performers shrink, in some cases to no more than an unrecognizable spot on the screen. At other moments, however, home viewers get extraordinary closeup views of performers — views they would never have as audience members attending a live performance. Thus, a matter as simple as the size of the video screen can violate the aesthetic of any production.

Browning also raises a different and rarely mentioned issue which affects the way a performance is translated to television. "Television today . . . has no authority. The viewer has more authority than the medium itself. There is nothing inexorable about it. You can look at it, turn away from it, turn it off, turn to another channel . . . it is so hard to get anybody to care about what's on television."

Television's lack of authority influences the way directors approach making television. "I unashamedly confess that I will do almost anything in my power to attract as wide an audience as possible. I will use almost any device I can to try to seduce the viewer to stay with the process," explains Browning. "The whole process is performer-oriented. What the audience responds to is not necessarily the legitimacy of the performance, but the energy in it. The key is energy. If you have a personality like Pavarotti, just get in close and

1. The comments by Kirk Browning, Fabrizio Milano, David Jones, Bret Adams, and Jay Broad are excerpted from *Cable Television and the Performing Arts*, ed. Kirsten Beck (New York University, School of the Arts, 1981).

The full cast of the Metropolitan Opera's production of *Idomeneo* (top), and the camera moving in on its star performer Pavarotti. (Photos: J. Heffernan, Courtesy of the Metropolitan Opera.)

show him. The chemistry is mesmerizing."

Fabrizio Milano, a Metropolitan Opera stage director whose work has also been televised, says that the primary problem for the director transferring a performance to television "is conceiving what the audience should see. The audience in the theatre chooses from a large stage what to focus on. The television director has to make this choice for the home audience."

Milano agrees that the entire spectrum of an opera cannot be captured on television. "What you *can* get," he emphasizes, "is the extraordinary energy of particular performers. Instead of trying to capture the expanse...you may focus on certain principals. Certain opera singers have fine acting technique, and it communicates on camera....the entire body, face, hands, and arms become an expression of the musical impulse. That excitement, intensity and energy can seduce an audience to stay with you...and can also compensate for not being able to give the full panorama."

Merrill Brockway, who was CBS Cable's executive producer, and prior to that the producer of WNET's "Dance in America" series, agrees with Browning that translation of a live performance from stage to television is not an art form. Brockway thinks it is fair to characterize this process as "just craft" when a live performance is taped. But, for Brockway, art enters the picture when a production is re-mounted for television. "When we take a work into a studio, it is the *aesthetics* of a piece that we are adjusting. It has to do with space and time and what the screen can and cannot do," he explains.

"Take Mr. Balanchine, for example," Brockway suggests. "He *hated* to tape from the stage because the spacing is different. He had been known to get those dancers up there in the studio and insist, "Millimeter, millimeter!" He *had* to recompose for the screen; otherwise the spacing would have been all wrong."

The process of transferring a performance to tape or film is problematical; few who have made the attempt will disagree. But the difficulties inherent in performance transfers pale when compared to the difficulties involved in representing the visual arts on television. Speaking at a 1982 Volunteer Lawyers for the Arts' conference on cable production, Wendy A. Stein of the Metropolitan Museum of Art's Office of Film and Television enumerates the difficulties this way:

> Art is incompatible with television. Art basically is still; television wants action. Art is silent; television is a chatterbox. Art comes in many sizes and many shapes. Television obliterates that; everything is the same size and it wants everything the same format as much as possible. Art requires time; TV wants speed. Art also invites you to choose a pattern of looking; film and television give that choice to the director. Art is made of variable colors; TV inevitably is going to distort [color] ...simply because it doesn't get the interactions of different colors in different lights. And finally, art is rare, it is unusual, frequently it's unique; TV will bring [art] into everybody's living room and make it universal.

Why Try Television?

If the prospects for artistic success are really so dismal, why try television? Money is usually the first reason suggested. Moving a work from its natural setting—be it a stage, museum, studio, or other location—onto tape usually involves substantial compromises, and it is rare for the new version to be as satisfactory as the original was. But money is often a powerful consideration in putting work on television.

Officials of Ballet West, the Salt Lake City-based dance company, were so excited when they realized that the company made money from their tapings for Bravo that "we were already planning to go into the business!" laughed company manager, Jane Hayes Andrew. "After all our expenses were paid, we still had money left over to go into general operating support." This happened at a time when Ballet West was looking for new sources of earned income. "You know, you look around and say, 'What can we do to raise money other than sell tickets?' It seemed so obvious. Television. Then I started checking around with all the other dance companies and found out that nobody else was really making money. Many were *raising* money to perform on PBS. That led us to conclude that television really was not as potentially profitable as it had appeared. It was just a one-time fluke for us."

More often, companies, artists, and performers turn to television for increased exposure. It is hard to quantify the results of television exposure, and there are risks involved. The product does not always turn out the way the artists hope it will. In one case, discussed later in this chapter, the program was so unsatisfactory that it was never cablecast.

Even worse, a tape that the artists consider totally unacceptable may be sent out for the world to view. There are, however, cases where a completed work has satisfied all who took part in the production.

Merrill Brockway is convinced that television exposure can affect a company's bottom line. "It became increasingly clear to me during my time with 'Dance in America,' for example, that appearing on that program affects bookings. It gives the company panache, celebrity. It's good promotion and gives the company stature in its community," says Brockway.

A study conducted in 1976 by the National Research Center for the Arts, Inc. for the Joffrey Ballet appears to confirm this as well. Audience members were questioned following their attendance at the Joffrey Ballet spring season. The company had appeared on "Dance in America" early in the season. Of first-time attendees, 59 percent felt that they had made the decision to attend the spring season "after" viewing the "Dance in America" program, while only 5 percent of the "regulars" reported deciding to attend "after" viewing the program. Although these results did not establish a causal relationship, the evidence points in that direction.

A more dramatic example comes from Curtis Davis, director of programming for ARTS. "About fifteen years ago, a nearly unknown Indian musician, Bismillah Kahn, was scheduled to make his first U.S. appearance. He was set to appear at Philharmonic Hall, and ticket sales were practically non-existent when PBS aired a segment featuring him. The film ran on Channel 13 in New York City ten days before the concert, and he was sold out. It was clear that, for the audience capable of being interested in such an artist, that was the

click, the moment of recognition. Suddenly, his name in *The New York Times* ad prompted a response: 'Hey, I just saw him on television; I *can't* miss that!'"

Philip Semark, president of the Joffrey Ballet, has found television exposure invaluable for his company. "There is no substitute for a live performance, and it is my impression that the television experience incites people to go to a live performance." When asked if putting the Joffrey Company on television was worth the artistic risks, he replied "Absolutely. It means more money for the company. Putting together a television program means more work for our dancers and staff, which supplements their salaries. By the same token, television increases audiences, which means more income at the box office. So there are benefits on both sides: more cash for the box office and increased fees, activity, and exposure for the dancers."

Another reason companies consider putting their work on television is for marketing and audience development purposes. David Jones, a theatre, television, and film director, tells an amusing story of a television feature becoming a marketing tool that backfired.

"When I worked at Monitor, the BBC television magazine, we would feature ten- to twelve-minute theatre extracts. Because we were purists, we insisted on removing it from the theater and putting it into the studio to re-pitch and re-gear it

Ballet West performing *Pipe Dreams*, which was successfully taped for cable by **BRAVO**. (Photo: © **Paul Kolnik.**)

for television. One of our most successful efforts was of
Zeffirelli's original stage version of *Romeo and Juliet*, where
most of the balcony scene took place in a large, fourposter
bed. We got marvelous camera work in, around, and under
the pillows, great closeups. Ticket sales went up enormously
the next day, but then people started to get angry because
they could not get as close to the bed as the camera had...."
That feature sold tickets, even though that was not the intent
of the piece.

Semark candidly admits that the Joffrey chooses its season
with an eye on what will appear on television during the season. "We kept *The Green Table* in our repertory this year, knowing that it would be shown in December and again as we go out
on national tour. We know that our actual audience size for
any program correlates with material shown on television."

Money, exposure, marketing, audience development: each
is a possible reason for wanting national television exposure
for a company. But are works ever put on television for artistic reasons? Yes, certainly in the case of Twyla Tharp.
Artistic reasons were also important in the decision to put *Gertrude Stein Gertrude Stein Gertrude Stein* on CBS Cable. As video
activity becomes more commonplace, more artists are using
it as an additional arena for their work.

Merrill Brockway has been observing artists' attitudes
toward television over a number of years. "When we first
started doing 'Dance in America,' we talked a lot about
translation," he says. "It took *years* for people to realize that
you couldn't create good television by taping the dance as it
is. There was so much *resistance* to television that they said at
first, 'Okay, but do it carefully, just as it is.' Then gradually,
they began to extend pieces. On *Short Stories*, for example, it
was Twyla's decision to go to close ups. For her, the piece
didn't have anything to do with feet."

Good arts programs are impossible to make for television
without directors and crews who understand and are sympathetic to the process. "Television directing is very complicated," says Brockway. It's like flying a jet plane." As
recently as 1981, David Jones found it difficult to locate
knowledgeable and sympathetic directors and crews in the
U.S. At the same time, he pointed out that for people in the
arts, it is not all that difficult to begin to understand the
televising process. "There is no mystery about it, you just
have to start spending time to learn what it's all about; begin
to get yourself into situations so you can begin to feel at
home."

Pat Carroll, whose performance in *Gertrude Stein Gertrude
Stein Gertrude Stein* was taped by CBS Cable, speaks almost

reverently of the camera crew (at Opryland in Nashville, Tennessee) that worked with her. "I have worked with television crews in New York and Los Angeles during my twenty-seven years in film and television, but I was in awe of that wondrous crew down there." She tells the story of the shooting of one particularly demanding scene, the one in which she speaks directly to her brother, Leo. "Ronnie Smith, the cameraman said to me: 'Now Pat, looka here. Ah'm gonna play Leo an' you just stick with me, 'cause ah'm gonna be they-uh fuh you.' And he was. He had taken the script home the night before we were to shoot that scene and read the scene over and over. I played to the camera, and he became Leo—that camera was Leo to me because of that wonderful young man."

Pat Carroll's gratifying experience with the Opryland camera crew has been repeated by others who have worked in the Opryland Studios. When David Jones speaks of his early experience in England, however, the story is different: "Although it's begun to fade now, there was for many years an emotional and artistic tension between the television and theatre worlds. On both sides of the fence, there was a disgraceful lack of awareness of just how remarkable the work being done on the other side of the fence is. The theatre people would say, 'It's those philistine television men, all they want is marvelous shots, and my production will go out the window.' The television people would say, 'Jesus, it's those artsy farts coming in here, and they're gonna be narcissistic, and they're gonna say that beautiful moment can't be lost, and it's all gonna be impossible.' "

Often, the truth lies somewhere in between, but, to listen to stories currently being told, the relationship between people in television and people in the arts still tends to be a little tense. This tension will begin to dissipate as there is more crossover between fields.

Although it is a daunting challenge to televise art without destroying it, cable's national cultural program services have managed to do so on a scale never before attempted in this country. Cable has given viewers more arts programming than ever before in the history of television. The quality of some of this programming is very high; some of it is terrible. But with continued opportunities to create new arts programming, quality will improve as people develop new and better techniques. Cable television, following the lead of PBS, has proven one thing to America: It is possible to make marvelous art on television.

The examples and experiences which follow grew out of programming created for national cable services. Subsequent chapters will detail cable programming on the local level and offer practical suggestions for putting the arts on cable television.

The Guthrie Theater Becomes a Television Producer

The Guthrie Theater has gone through a long series of television experiences and has taken on increased producing responsibilities with each project. The most recent, *A Christmas Carol*, was produced solely by the Guthrie, whereas *Camille*, the company's first taped production, was produced by an independent producer and received mixed reviews.

Bret Adams, a New York City-based agent, contends that *Camille* exemplifies why productions should not be taped live. "I saw *Camille* live," he said, "and it was a truly lovely, delicate show. The outfit that produced it for television didn't allow enough time to do it properly, so you have this lovely, delicate production which you cannot *see*, because the lighting was not right. The lighting was not re-done, so the color is all wrong. The [television] production appears purple, deep blue, almost black." Guthrie Managing Director Donald Schoenbaum agrees that time was a problem, but disagrees with Adams' estimation of the tape: "The product may make some artistic directors cringe, but I don't think it's that bad."

The next television project undertaken by the Guthrie was *Vincent*, Leonard Nimoy's one-man play about Vincent Van Gogh, a joint production of the Guthrie, ABC Video Enterprises, and Magnavox Productions. "The executive producer of *Vincent* was on the Guthrie staff, but the theatre didn't have the total financial responsibility for the production as we did with *A Christmas Carol*," explains Schoenbaum. The Guthrie was fully satisfied with the quality of the *Vincent* tape and after that was ready to do more television.

Margaret Whitton as Marguerite in the Guthrie Theater's production of *Camille*. (Photo: Bruce Goldstein.)

A Christmas Carol became the first television production for which the Guthrie took full producing responsibility. This was partly because The Entertainment Channel (TEC), which had approached the company about cablecasting the show, did not produce its own programs. Rather than have TEC hire an outside producer, Schoenbaum decided that the Guthrie, itself, should produce the tape. "I felt I could control it better than anyone else could—both budget and time—and that proved to be true," said Schoenbaum. "Also, I felt very secure with the show. I know it very well, the staff does, and so do our technical people. We had done it for six years at the time of the taping."

Schoenbaum was aware that there were risks involved, but he felt that the artistic risks were the least of them. The key to making the deal work was being able to do the entire taping in thirty-two hours—one twelve- and two ten-hour days. The taping took place on a Sunday, Monday, and Tuesday, with regularly scheduled performances going on at night. If the taping could not be finished in that three-day span, substantial expenses would have been incurred to hold everyone over until Thursday, because Wednesday was a matinee day on which there could be no taping. "By producing it ourselves, we were actually protecting our artistic product," said Schoenbaum. "If it came to a need for more money because we had to hold over for additional taping, I would be dealing

Leonard Nimoy in his one-man show *Vincent: The Story of a Hero*, the story of painter Vincent Van Gogh (top), (Photo: Courtesy of The Guthrie Theater and Leonard Nimoy), and The Guthrie Theater production of the Dickens classic *A Christmas Carol*. (Photo: Courtesy of The Entertainment Channel.)

with my own money instead of someone else's.

"If your organization has the opportunity to be executive producer for television, make sure you have the resources to do it — the key people are the producer and director. The fact that Bill Siegler was available and willing to serve as producer also influenced my decision to serve as executive producer," said Schoenbaum. With the Guthrie's *A Christmas Carol*, careful planning, knowledge of the production, seasoned performers and staff, and good television people came together to create a very satisfactory television program.

Going Over Budget

The taping of another company's production of *A Christmas Carol* was not nearly as well planned as the Guthrie's. James B. McKenzie, executive producer for American Conservatory Theatre (ACT) in San Francisco, offers the following advice to companies involved in cable projects: "Budgeting and planning are most important for a project. You want people you can really trust who understand both art forms. Don't proceed until everyone has approved every detail of the shoot, not just the money, but the technique, the style, the ideas, everything. I think you have to do the film on the desk before you move into the studio." This wisdom springs from McKenzie's reflections on ACT's experiences taping *A Christmas Carol* for ARTS, an experience he describes as "a nightmare."

The production began with an approved budget of $200,000, but ended up costing ABC Video Enterprises $400,000. How could that happen? As McKenzie explains it, the original budget was set without an adequate understanding of the concept of the production. "The directors, both stage and television, had a concept different from the budgeters, and they weren't able to talk to each other enough to figure that out, partly because of the time considerations. Both parties agreed to do the show at too late a date. We should have had five or six months after we signed the contract for pre-production planning, but as it turned out we were really caught. When we finally signed the contract, we had only thirty days to get the show done, because we had to be out of the theatre where we were taping before ACT's season began."

When asked to comment on McKenzie's diagnosis of the problems with *A Christmas Carol*, Mary Ann Tighe, vice president for program development at ABC Video Enterprises, agreed that the primary problems were time and lack of proper planning. "We learned our lessons on that particular play," she said. "It was the first theatrical show we ever did, and there was no programming staff at the time, so there was in-

sufficient coverage from the ABC side." "It was a good experience," she concluded, "because we've *never* let it happen again. . . . from that time on, we've never gone one penny over budget on another theatrical work."

McKenzie is philosophical about the experience and, having seen the final product, is satisfied. "Basically, I think we're both very proud of it artistically," he concluded. Those at ARTS concur.

An Artistic Failure

Bad planning can have a disastrous effect on the budget, the experience of taping, and on the artistic product. One of the first dance tapes produced by CBS Cable was Kenneth Rinker's *40 Second 42nd Variations*. Rinker was approached by CBS Cable and asked to re-mount the work, which he did. It was taped in Nashville at the Opry studios as was much of the dance CBS Cable produced. Although the piece was fully produced, CBS Cable never used it.

According to Rinker, the piece was never cablecast because "it didn't make any sense." "I see the piece," he says, "as an abstract movement piece with a kind of narrative subplot." Rinker thinks the CBS Cable people saw the work the other way around, as being a narrative work with abstract elements. "From pre-production all the way through editing, it seemed as though the piece was being shoved more and

The cast and crew of The American Conservatory Theatre and ABC Video Enterprises production of *A Christmas Carol* (top), and the director and crew looking on during taping. (Photos: Larry Merkle, Courtesy of The American Conservatory Theatre.)

more in the direction of strict, straightforward narrative, to the point where there was nothing left to the imagination. So it ended up being totally graphic," he continues. "When it's disco, we're gonna do flashing lights, and when it's blues, we're gonna tell you 'bar' and when we're on the street, we're gonna say 'street light' with actual props." In giving this explanation of how the piece went wrong, Rinker is not placing all the blame on CBS Cable: "Part of this was my doing," he concedes.

By the time the piece had reached the editing process, Rinker had become aware that it was not working right. "I had to stop — to rethink what I had done originally, to go back to the piece itself. And when I did that I realized it didn't make any sense to try to give it more narrative clarity. It didn't have that in the first place; that was something that was shoved on top of it." Ultimately, in Rinker's view, it was the unresolved tension between the narrative and abstract elements in the work that caused it to fail artistically.

But there were other complicating problems. The original schedule called for three days of shooting. Because of problems with the set, the first half day was lost to set changes, and they ended up with only two and a half days of shooting time. But not only was time lost, the shooting schedule for the piece was determined by the logistics of moving the set. "We had set pieces that could only be moved in a certain order logistically and economically, so that order determined the order in

The Kenneth Rinker Dance Company performing *40 Second 42nd Variations*. (Photo: Otto Berk.)

which we were taped. The piece wasn't taped from A to Z; it was taped according to the movements of the set. This was still being determined even as we got to Nashville, although some was known before. The set builders were, in a way, determining the order of our dancing—by what was ready, what color the floor was going to be, and how long it would take to dry once it was repainted. Some of those things were known beforehand, and some of them we did not know until the moment that we did them."

The order in which the piece was shot caused the quality of the dancing to suffer, Rinker feels, because the section that demanded the strongest performing was shot last, when the dancers were tired. Ironically, that order was determined to a great extent by a set designed with the narrative elements of the piece in mind. So, in one sense, all the difficulties experienced with the Rinker piece can be traced to the need for a clear vision of the work as a television piece.

Is there a way that the situation could have turned out more favorably for all concerned? Rinker feels that two changes would have gone a long way toward making his situation more favorable. First, he cautions everyone entering a business relationship in order to put a work on tape to be clear about the relationship that is being established. "If you can look at it from the beginning as a game, do so," he says. "Determine what the rules and regulations are, what the size of the court is, the number of players, the type of equipment—determine everything you can think of in terms of facts. That will make it much easier when it comes to dealing with the vague kinds of questions of how you decide what to compromise artistically." Finally, Rinker says, you can only complain if you asked for something, it was not given, and the result was bad. If you do not ask, you have no right to complain. So he encourages companies: "Don't be afraid to ask for what you need. If you need a large studio space, if you need pre-production time, distance on your work, if you want somebody to come in and videotape so you can get an idea of how you want to put something together, then ask for that. Maybe they don't want to give you these things. Maybe they can't. But it will help you in the long run to have these things." Rinker says anything that helps people attain clarity in their aims and their business relationships with their producers is useful. No one who has been through the transfer of a work to television would disagree with that.

In summing up his experiences with CBS Cable, Rinker says, "I guess I came away with not so much sour grapes as the conviction that dance doesn't really work so well in the television medium. I think the video experience teaches you something about your own work. Personally speaking, I think the most beneficial thing about it is that it brings me back to my own work as a choreographer. I do not want to be a video director, and if I do another video project, I would want to collaborate closely with the director, but I would prefer that the director have total control over the directing so that I can sit back and be more objective about it."

Bringing the Camera *Inside* Dance

The video and film experiences of Alison Becker Chase have led her in a very different direction. As a choreographer-performer and co-founder, with Moses Pendleton, of Pilobolus, she has experienced several transfers of the company's work to television, and has become increasingly interested in exploring the process herself. "I don't think we've begun to do the filmic translations we are capable of doing."

Chase finds her experiences with film and video seductive — not so much because she considers the work done thus far so artistically successful — but because the work done to date points to new directions for exploration. "The audience fantasy," she says, "is to be right up there while the performance is taking place. It's so exciting to be right in the middle of it! And the camera can show things the audience never sees. But so far, they always seem to place the cameras out there in the audience."

Chase considers most examples of Pilobolus on television unsatisfactory. "We have had a great deal of exposure, but always seem to have to conform to 'the Barbie Doll' presentation of modern dance on television. Our work, for example, has a certain amount of nudity in it, but that's not allowed." It is as though the company's performance is cleaned up, squared off, and shot from the perspective of the audience with no attempt being made to capture any of its special personality. "After these experiences," Chase says, "we have always ended up feeling that we have lost control of our imagery — that the transfers were just frontal documentation."

The one exception is *Moses Pendleton Presents Moses Pendleton*, produced for ABC Video Enterprises by Fort Worth Productions. For the first time, Chase says, the company was not simply "turned into dancing bears." "In *Moses Pendleton Presents Moses Pendleton* the camera leaves the proscenium and really presents Moses' imagery. It shows what he saw when he took a shower and how he sees his house." The film is a sixty-minute documentary which includes a generous amount of performance footage. The final scene is a dance specially choreographed in a candlelit setting inside Pendleton's Victorian house. The work is performed by Pendleton, Chase, and Daniel Ezralow, and the choreography, the camera work and the setting are stunning. "The scene with the candles was such fun to do!" exclaims Chase, "it wasn't just trying to duplicate the proscenium situation; we made a new environment to dance in. It was an image that Moses and Elfstrom [Robert Elfstrom, the director] had been evolving over the course of the film."

Chase traces the success of *Moses Pendleton Presents Moses Pendleton* to the relationship between Pendleton and the film's director. "We had worked with these people before, and Elfstrom and Moses had connected then. It continued to be a successful pairing. Bob would say, 'Come on, look at this image,' and Moses would look and say, 'Oh yeah! I like that. Couldn't we go even closer?' And that's the way they worked."

"The ABC project allowed for much more impromptu footage," Chase adds, "whereas the other experiences were so methodically planned that they couldn't take off on something that was happening at the moment. Although the candlelit scene had been thought of before, and the film had been scripted, it was all very, very loose. That's the way Moses and the rest of us work best. The script was just sort of a hinge to get a go-ahead to do the film. It was always being re-written and re-thought as they did it and as they looked at the rushes."

The experience with *Moses Pendleton Presents Moses Pendleton* has Chase excited about continuing the process. "I want to bring the camera inside dance." The company might want to take an old piece that has already been choreographed, for example, and "really go in there and get exciting footage and re-edit it as a specific dance for film and then put the music to it. What was the beginning of the proscenium piece might then become the middle of the film piece. You could really scramble it up and rephrase it. You could re-see it for film."

Bravo Does the Brandenburgs

Gerard Schwarz, music director for New York City's 92nd Street Y Chamber Orchestra, calls his orchestra's taping of the Brandenburg Concerti for Bravo in December 1981 a "one hundred percent good experience." Schwarz knows what he is talking about; he has been taped in studio and live concerts many times.

The Brandenburgs, an annual event at the Y, were to be taped during performances with an audience in the hall. "I had a meeting with the Bravo people about a month in advance and told them all the things I didn't want to happen during our taping—all of this based on my previous experiences with other companies. I said I didn't want anyone moving around with a large camera on stage during the performance. It's too distracting to the musicians, the conductor, and the audience. No talking. So many times when I've done commercial television work, the cameramen are talking to the director while we're trying to make music, and even though it's quite soft, it's very distracting. I didn't want cameramen on stage in blue jeans while everyone else was wearing tails. I also asked them to tape more than one program and to allow me to be included in the editing.

"As a result of our early conversations and understanding, Phil Byrd, the director for Bravo, was very careful not to interfere with the musical performance. We did our concert with some additional lighting, so it was a little hotter and a little brighter than normal. There was one cameraman who, at times, was on stage, but he was dressed in concert attire (a tuxedo). He was very quiet and didn't move around a lot. All we knew was that there was one 'foreigner' on stage who blended in very well, the lights were a little hot, and that was all. Bravo was very sensitive to the needs of the performers."

BRAVO cameramen in tuxedos (rear), taping a performance of the Cleveland Orchestra. (Photo: Peter Hastings, Courtesy of BRAVO, a Rainbow Service.)

How Performers are Affected

Listening to stage performers talk about their experiences being taped is enough to cause any manager contemplating signing a television contract to pause. The transition is particularly difficult for dancers, as Kenneth Rinker made clear when he spoke of how the performance quality of his piece suffered when the most demanding portions were scheduled to be shot at the end of the longest work day in Nashville.

Rinker speaks eloquently about the difficulty of adjusting to the rhythm of taping: "When you get ready to perform live, you do everything you need to do to be sure that when you go out on stage you are physically at your optimum to do the best you can do. Then you perform. When you tape something, you get yourself ready the same way — your rituals, your makeup, your warm-up — and then maybe you go on and maybe you don't. There may be a technical failure, a union break, a set change, or any number of other interruptions. Your day starts early in the morning, at six or seven, and can continue late into the night if you go into overtime. Most dancers are not used to working this way."

For a company used to performing live, the switch to taping conditions can be traumatic. It is worth expending every effort to make the dancers' situation as comfortable as possible. Such efforts will pay off in the quality of performance on tape or film.

Both Alison Becker Chase and Kenneth Rinker believe that television cannot convey the energy of the dance. As a result, Rinker says, "In television tapings you don't have to use energy like you do on stage, because it doesn't read very well. You just have to be clear about what you're doing with your body and where you're going. Don't do anything more or less than that. Too much energy is just wasted energy. Nobody sees it; nobody feels it. When you're on stage people in the audience experience it, but it doesn't transfer to television."

Chase talks about how difficult it is to break a work into pieces for taping. "If you are doing something kinetic, it's very hard to get that energy going when you do it piecemeal. And then the energy that *is* there doesn't translate to film or video." However, once again Chase said the experience on *Moses Pendleton Presents Moses Pendleton* was different from other tapings she had been involved with. "We had much more control of the performing situation than we did in our other television experiences. The crew was smaller and the experience much more intimate. If we were not ready when the cameraman said 'ready to roll,' we could wait or do it again. For example, we did a section of *Shizen* six times until we got

Moses Pendleton and Alison Becker
Chase performing *Shizen*. (Photos:
Joe McNally, Courtesy of Hearst/
ABC ARTS.)

it. Sometimes when you are working with a big crew, you can't afford to do it that many times." In that piece and in Moses' *Momix*, a great deal of the energy does come across successfully.

From the musician's point of view, being filmed or taped is a somewhat less daunting experience than it is for dancers and actors. James Barbagallo, an American pianist who won a Bronze Medal in the 1982 Tchaikovsky Competition in Moscow, was one of the performers featured in a documentary on the competition. (The documentary debuted on ARTS in November 1982 and aired some months later on PBS affiliate stations.) Barbagallo's first television experience was on cable television in the Bay Area of San Francisco in 1972. For that and subsequent tapings, he found that the anticipation of being taped increased his nervousness. "I am most affected just before I go out to perform," he says, "I think to myself, 'Oh, my god, there are all those mikes and cameras...' " But it was different in Russia at the competition. "The nerves of competition were such that I didn't even think about being recorded. When I would watch and listen to the other performers, I was very conscious of all the television equipment, but when I was playing, I was not. I was so unconscious about being taped that when I saw segments of my performance on Russian television, I was surprised. It was only then that I remembered that my performance had been taped."

Barbagallo says that watching the tape of his competition performance has been helpful in an unexpected way. "I have amended my playing style somewhat — changed my decorum at the piano. I think I moved around too much and showed too much of the music in my face. A couple of the judges commented on this to me in Russia, but I didn't know what they were talking about, not ever having seen myself perform. Now having seen it, I can see how my style might be distracting. So I'm going to make some changes, and that all comes from having seen the tape."

Fabrizio Milano tells how the Metropolitan Opera's approach to performers evolved: "When we began, we decided we would not disturb the performers at all — that they would not even be aware of the cameras. Now we are involving performers more and more. We'll tell the people they are going to be on a particular camera, so they know what camera to play to, and they are learning to trust the camera. For instance, Renata Scotto, when she saw the playback of herself, immediately decided that she was going to have to look different. She lost a great deal of weight and she looks tremendously better now. She has learned to control and scale her gestures so they will read well on the camera, so they do not look overblown. The more we work in this area the better we all get at it."

Alison Becker Chase found that it took time to get used to performing in front of the camera. "On the ABC Video film, I felt a little jealous of Moses, because he'd been doing so much more work in front of the camera and was much more impromptu and relaxed than I was. I wished I'd had the time to get at ease and playful with it. It feels very strange having this one lense-eye looking at you. As a performer you are used to a 360-degree focus, but with the camera you have a tendency to sort of square yourself off like you would with a proscenium when you really don't need to. There is a real art to performing for the camera which can't be developed except by doing it.

"At the end of the ABC film," Chase continued, "Moses was beginning to develop a relationship with the camera similar to what Pat Carroll had with the cameraperson who became Leo in Nashville. Moses began to see the camera as another person and he played to it as such; it wasn't a still response. The camera can be a blank, impersonal eye, or it can take a personal point of view. If you, as a performer, know you are playing to a personal point of view, it really changes your performing attitude."

"Everyone Just *Stop!*"

The decision to transfer a work from stage to television is a decision to take a risk, and the risk is not confined to the performers. Jay Broad, a playwright, tells one of the quintessential stories of the helplessness of a playwright who is not in control. A musical written by Broad had been shot in a studio with a limited budget, and Broad says that during this project he "found out how people who are not in control get such reputations for being insane, crazy, and also very silly."

Broad was in the editing booth, "as time was getting tighter and tighter, when the director said, 'We have to shoot that shot over,' and the producer said, 'No, we'll just have to go with it. We don't have time.' This had happened ten or twelve times that day, so I said, 'No, if we can't do the thing right, let's just stop the production. Everyone just stop!' And because someone in the booth had said 'stop,' everyone backed away from the controls. The producer said, quite rightfully, 'This man does not have the right to stop this production.' And she was absolutely right, so everyone went back to the controls. I finally thought to myself, 'The only way I can stop this is by throwing myself across the monitors, so they won't be able to see to continue.' At that point, I said to myself,

'That's insane!' So I left, and they went on. This is not critical of them, as it all turned out pretty well, but just to say you must understand clearly who is in control. If you cannot retain control, don't invest your soul in the project."

Gertrude Stein Gertrude Stein Gertrude Stein: A Happy Transfer

Although they did not retain complete control over the product, the steps taken by the team of Pat Carroll and Mary Ellyn Devery to protect the artistic integrity and performance quality of *Gertrude Stein Gertrude Stein Gertrude Stein* were extraordinary. It is hard to find anyone who has a bad word to say about the CBS Cable production of the play. Even the most hardened critics of theatre on television are complimentary about the production. From start to finish, the Gertrude Stein experience stands as a happy reminder of how well CBS Cable, a dedicated performer, and caring producers and administrators could work together to make a stage presentation into a first-rate television program.

Knowing they wanted to get the piece on tape, Carroll and Devery decided in favor of cable exposure for two reasons. First, they felt that with the predicted growth of cable television, there would be an increasing audience for the work for a period of years. Second, they wanted to retain more control over the final product than is possible in a network television deal. "We didn't want to cope with the demand for an 'Alice character' and dancing girls," comments Devery.

Although they were approached by a number of people, they went with CBS Cable because, as Carroll explains it, "Merrill Brockway's ideas seemed closest to what we wanted done. We felt that with Merrill, who had such a passion for the work, we were as safe as we could be." A key consideration for Carroll was that the play be taped without an audience, and Brockway agreed.

The negotiations took well over a year. Devery, the show's producer since its inception, was uncompromising in her determination to protect both the property and the performer. "There were certain points," she explains, "that Pat and I had outlined, and if any were missing, we were prepared to lose the deal. One of these was a provision that we share in the proceeds of any future sale of the piece."

Another key point was mutual agreement on the final edit. By contractual agreement, CBS could not cablecast the production until both parties had signed off on it—an unusual provision, Devery agrees. "I wore two hats. I would attend the creative discussions, and then I would be the only one from that group present at the negotiating, legal discussions."

During the negotiation period, Carroll and Devery decided on the "support team" they felt necessary for the best possible taped performance. The presence of this team—Devery, Milton Moss (the stage director), Dr. Julia Wing (Carroll's voice coach), and Ellen Zalk (the production manager for the stage version)—was negotiated into the contract. Carroll compares this to "having good people in the corner during a prizefight. They know when to pour water on you and when to give you a drink.

"Between takes I would sit on a lounge chair, and Dr. Julia worked on me, either doing my shoulders or giving me exercises to keep my diction going or to keep my energy up. Milton would come over occasionally and give me a point about characterization. Mary Ellyn was always there to ask if I was pleased with the last take, and Ellen was there on book and watching blocking. Everybody protected me, including the CBS Cable people," says Carroll.

Pat Carroll in the stage production of her one-woman show *Gertrude Stein Gertrude Stein Gertrude Stein.* (Photo: Gerry Goodstein.)

Did this kind of care prevent dissatisfaction with the final product? For the most part, yes, but there were problems. The set design was one major disappointment. During the only meeting Carroll and her producer had with the designer, they felt he had already made up his mind about the set concept and consequently didn't seem to hear what they wanted. When they saw the set model the first day of rehearsal, they were not pleased. Their solution was to divert attention from the set by keeping focus tight on Carroll.

Another disappointment was the first cut. After seeing it, Carroll says, "I was ready to get up, walk out, and throw myself in front of a bus, but Mary Ellyn said to me, 'No, sit here and watch it, and make notes. How do you expect these people to know if there is displeasure if you don't make notes?' " Thanks to the contractual agreement requiring a mutual okay on the final edit, CBS Cable did make the changes Carroll and Devery wanted, and the final result pleased both parties.

Pat Carroll agreed to put *Gertrude Stein Gertrude Stein Gertrude Stein* on tape because it represented the final step in the development of a property for which she felt live performance was the most important identity. "It was a risk, but it had to be done at some point, because until it is put on tape, where it can be saved, archived, or burned, I have not paid my full debt to everyone involved in the project, including myself. Having done that, I feel there are certain moments of the play that come forward as they never have on stage. Of course, I haven't seen it on the stage because I'm involved in it. But I do think, sitting in front of that television set and watching it,

The full setting of the stage produc-
tion of *Gertrude Stein Gertrude Stein
Gertrude Stein*. (Photo: Kirsten
Beck.)

there were moments when I absolutely became an audience. I
was not that person performing. I was watching this woman
involved in something, and she captured me."

It would be a mistake of serious proportion, however, to
conclude from the foregoing tales that the only way an arts
organization can have satisfactory television exposure is
through a national medium such as ARTS, Bravo, or PBS.
There is another chapter in the story of cable television, and it
may end up being the most important chapter as far as the
arts are concerned.

Cable's Access Ability

Television is the dominant medium of our time. Again and again, studies have demonstrated that the vast majority of Americans get most of their news and information (not to mention entertainment) via the tube. Yet, as television has grown ever more powerful in its influence, the ability of the individual citizen to have any appreciable influence on the programming broadcast over the airwaves has nearly vanished.

As in the broadcast industry, the cable industry at the national level is dominated by huge communications conglomerates. But in most cases, cable *operators* are regulated locally by municipalities which hold the rights to grant franchises. (Several states have statewide franchising authorities.) This municipal control results from the cable operator's use of public rights of way to string cable on utility poles or to lay it underground. The terms and conditions under which a municipality grants permission to use these rights of way are contained in the franchise agreement, which specifies the franchise fee the cable operator will pay and stipulates the kind of telecommunications system the company is obligated to provide. Granting a franchise is the ultimate exercise in local determination of communications policy, and the care with which the agreement is drawn up determines the level of enforcement a community will be able to maintain.

Because of cable's potential as a delivery system for not only entertainment but other electronic services as well, the stakes are high in the franchising process for the cities, the citizens, and the cable companies. Cable companies are competing for franchises because of cable's potential power as a money-maker. Cities are charged with the responsibility to negotiate the most useful and advanced cable system they can secure for their citizens. Citizens with outdated cable equipment and systems will be left behind as new services (banking at home, shopping at home, home security, etc.) become more widely available. Information transmission via cable is expected to become commonplace in the future.

Cable television franchising is a political act and has all the pluses and minuses of local politics. Fortunately, there are some significant pluses, not the least of which is the fact that

Cable franchising is an issue on which citizens can speak and be heard.

franchises are granted by local elected officials, who are responsible to the voters who elect them. This means that the people who use cable are much closer to those who regulate it than they are to the FCC commissioners in Washington who regulate the broadcast industry. Consequently, cable franchising is an issue on which citizens can speak and be heard. This has implications for the community at large and for the arts community in particular. If enough people speak with one voice, political officials will listen.

Citizens with a special interest in particular aspects of cable service who make an effort to be heard during the franchising process can have a genuine impact on the outcome. Arts, educational, and other nonprofit organizations in particular are in a position to benefit from the negotiation of a good franchise, but city negotiators must be properly informed and cultivated. For arts organizations seeking access to a locally run system, it is wise to take an active role in the franchising process from the beginning.

Franchising Wars

Local regulation has other implications which are not so favorable for communities. The rush to secure franchises in the cities and suburbs has spawned "franchise wars" among the cable companies bidding against each other. According to Les Brown, the editor of *Channels*, "To win one of the big plums . . . you simply promise the moon and pay no attention to whether it fits into the business plan." Brown describes an incident that reportedly took place in Los Angeles:

> "After each of the companies made its pitch for the franchise, a city official asked if any one of them, having heard the others, wished to sweeten his bid. The representative of one company immediately augmented his proposal by ticking off a list of satellite program services that seemed to provide for every conceivable interest. 'Hey,' one of his rivals protested, 'he's just bid two non-existent services.' Another jumped to his feet. 'I bid *four* nonexistent services!' "

In a pioneering series of articles in *The Nation* in the early seventies (later published as a book, *The Wired Nation*), Ralph Lee Smith made a case for cable television to be treated and regulated as a common carrier. But in a follow-up, published in *Channels* in 1981, he noted that although his ideas had attracted widespread attention, no action had ever been taken. Observing the franchising wars of the late seventies and early eighties, Smith once again made a plea for the creation of cable policy at the federal level. He cited the free-for-all of bidding and observed that even those who would be expected to serve as a steadying force in such a volatile situation have not done so:

> "By joining — and, in fact, leading — the every-man-for-himself melee that now passes for national policy on cable, groups wearing the mantle of public interest look more like part of the difficulty than part of the solution. The essential problem is the absence of any national policy on cable."

If franchise bids have entered the realm of fantasy, then when it comes time for actual delivery, no community can know what to expect. Still, although it is highly unlikely that the more extravagant promises will ever be kept, cities and towns and the special constituencies within them do not feel comfortable scaling down their franchising demands when communities all around them are being promised the moon and the sky.

The time of reckoning is upon us, and the cable industry and interested parties are doing what they can to hasten it along. As the building of new urban franchises proceeds, cities are trotting out the enforcement provisions negotiated in their franchise agreements, trying to hold cable operators to their construction schedules and their system designs. At the same time, the jurisdiction of municipal governments over cable companies is being challenged in the courts, and bills have been introduced in Congress to strip the cities of their regulatory power. Despite repeated efforts by the cable industry to wrest regulatory control from the cities and states, franchising and subsequent oversight still remain local prerogatives. But the industry continues to fight this battle.

To further complicate matters, one legal challenge has had a drastic effect on at least one city. In Denver, Colorado, Mile Hi Cablevision had won the franchise and construction was about to begin in mid-October 1982 when the Mountain States Legal Foundation, a nonprofit organization formerly headed by James Watt, filed suit against Denver and the cable company. The suit alleges that the first amendment rights of Denver's citizens were violated by the city's grant of *de facto* exclusive franchise to a cable company.

Not only does the suit seek to undermine the city's legal right to grant a franchise, it also puts Denver's franchise in legal limbo. With the Foundation's announced intention to pursue the case to the Supreme Court if necessary, the final outcome could take years to determine.

Consequently, Mile Hi's lenders withdrew a portion of the financing originally committed for the system's construction. This resulted in a redesign of the system and a cutback of

> "...groups wearing the mantle of public interest look more like part of the difficulty than part of the solution. The essential problem is the absence of any national policy on cable."
> —Ralph Lee Smith

community service. Although construction began in February 1983, the system being built is not the same one the city originally negotiated.

Yet, amid all the confusion created by these changing conditions, cities and towns continue to grant franchises. Those who wait for the ambiguity to clear may never have a chance to influence the shape of their telecommunications system. Under these circumstances, a community's only real choice is to determine what its citizens need and can use in a cable system and set out to find a company that will build and operate such a system.

A Cable System Can Benefit the Arts

In some communities, the advent of the cable system has brought direct benefit to the arts in the form of special channels reserved for the use of the arts community. In some cases, funds to help with programming costs for these channels have been promised as well.

Sometimes arts organizations have received direct benefit from cable companies. For example, as part of the franchise agreement for La Mirada, California, the La Mirada Civic Theatre was equipped with additional power in order to serve as a taping facility. The cable company, California Cablesystems, a subsidiary of Toronto-based Rogers Cablesystems, also established a $700,000 revolving capital fund for production expenses and provided a $50,000 fund for consulting and legal expenses to create cable programming within the theatre. In Lafayette, Louisiana, as a part of the franchise renewal agreement negotiated with Lafayette Cable Television, an access organization was established which is housed in the arts council building and receives operating support from the cable company's revenues and the city of Lafayette, Louisiana. Four channels have been designated for public access, of which one is currently in use.

The local nature of cable television affords artists and arts organizations in most communities the opportunity to make a significant impact on the kind of cable system to be built and how it is to be operated. For this to happen, however, the arts community must understand the process by which cities grant franchises to cable companies and the ways in which franchise agreements can be written to benefit the arts. Involvement in the process must not end with the signing of the agreement; the arts community must also be active in the process of enforcing fulfillment of the cable operator's commitments.

Access: What It Is and Why It Is Important

Access is the guaranteed right to use the local cable television system. *Public* access guarantees individual citizens the right to use a certain amount of time on the cable television system for programming they have created. The content cannot be controlled by the cable operator. In many cases, entire channels are set aside for certain kinds of access. For example, East Lansing, Michigan has six access channels: one for the general public, one for the public schools, one for city government, one for the public library, and two for Michigan State University. New York City's borough of Manhattan has public, municipal, educational, and leased-access channels.

Broadcast television is subject to the FCC's fairness doctrine and equal time rule. The rationale for both regulations is that the spectrum space used by broadcasters is scarce (therefore precious) and the airways used by broadcasters actually belong to the public, so the public interest must be preserved in their use. On the other hand, cable television is neither a medium of scarcity nor does it use the public airwaves. Consequently, it is not subject to the fairness and equal time rules.

Access channels accomplish for cable television a much broader version of what fairness and equal time were intended to accomplish on broadcast television. Access gives a television voice to the dissenter, the unpopular, and the minority as well as to organizations working in the public sector and in the public interest. Access is a video voice for the people.

Access began in 1972 with the FCC's Second Report and

La Mirada Civic Theatre, equipped with additional power in order to serve as a taping facility. (Photo: Gil Jordan, Courtesy of La Mirada Civic Theatre.)

**Access is a video voice
for the people.**

Order, which required major-market cable systems to have a twenty-channel capacity and to set aside one channel each for educational, municipal, and public access programming. Systems with 3,500 subscribers or more were also required to provide production facilities for local access programs. These regulations were overturned in 1979, when the Supreme Court ruled that the FCC did not have the authority to regulate channel capacity or to require access channels. While some of what was once required by the FCC has been negotiated into individual franchise agreements, the right of the average citizen to make and cablecast a television program without charge for air time is no longer guaranteed by FCC regulations.

In 1972 there was no satellite-delivered programming, and some operators with twenty channels to fill welcomed the dedication of three channels to municipal, educational, and public use. There were few other low-cost ways to fill the channels. But today, a twenty-channel system does not have a channel to spare. Furthermore, the current abundance of pay services has made channel space extremely valuable. *Village Voice* cable columnist Bob Brewin wrote in October 1982 that the worth of a channel on Manhattan Cable to a pay-service supplier — if only half of the system's subscribers take the service — was estimated at $850,000 a month.

The loss of the FCC mandate, the proliferation of satellite-delivered programming — both basic and pay — and a general trend away from regulation have combined to create a rough road for access channels and their users. Most cable operators would rather see their electronic real estate being used to earn more money, and they resent any outside interference in their programming practices. The National Cable Television Association (NCTA) contends that telling a cable operator what to put on the system's channels is like telling a newspaper what to publish. The argument goes as follows: Cable operators are electronic publishers, and any municipal requirement of access (including the franchise agreement between the city and operator) constitutes government regulation of program content, and government regulation is a violation of the first amendment.

George Stoney, professor of film and television at New York University and a founder of the National Federation of Local Cable Programmers (NFLCP) suggests that the country may eventually need an "access rights amendment." He points out that less and less of our population votes in each election. Since most people receive most of their information from the electronic media, television could be used to get more citizens involved in the process of government. If people are to be involved and educated, there must be access to them through the media.

"I see cable as the rehearsal ground for all the electronic media," says Stoney. "We must have access not only to cable, but to on-air television and on-air radio. Some may counter that there is not enough space, but it is all a matter of perspective. We are talking about the freedom of speech and the survival of our democratic government."

> "We are talking about the freedom of speech and the survival of our democratic government."
> — George Stoney

"It's interesting," Stoney observed in *Access* in July 1981, "that a basic copyright law was a part of the American Constitution as adopted. But the First Amendment, which guarantees us the right to say what we please with our art, had to be added later. Surely, we can see there is need to extend this concept to the electronic media. In the Reagan regime, it may seem ridiculous to suggest that we need a new constitutional amendment that will guarantee the public's access to all electronic media, but I think that in time the good sense of this statement will be recognized."

For arts organizations, too, access can provide a means of reaching the public via the electronic media. Traditionally, lack of money has barred the arts community from using television, society's most potent promotional tool. But access, as it works in some cable systems today, overcomes many of the financial barriers and allows the arts to enter the video age.

Access, Local Origination, and Community Programming Defined

Broadcast television has long been castigated for its lack of attention to local news and public affairs. One characteristic that sets cable apart from broadcast television is the relative ease with which a cable system can create programs for its community. Many cable systems have stepped into the void left by broadcasters and have begun very successful programs that cater to their subscribers' local concerns — from cablecasting city council hearings and local high school sports to full evening news programs devoted entirely to the local scene. Additionally, as cable companies are forced to compete with alternate delivery technologies (direct broadcast satellites and satellite master antenna systems, for example), local programs are a service cable can offer that other technologies can not.

To take advantage of local programming opportunities, the arts community has to understand the rules of the game. The first step is to grasp the important difference between local origination and access, the two traditional methods of producing local programming within cable systems. In local origination programming the operator has total control over the program, from initiation to cablecasting; access programs are originated and created outside of the operator's control and are run on the system at no charge to the producer.

Local origination programming (also called *local o* or *L.O. programming*) is produced by staff in the company studio, using company equipment. Almost all cable companies seek advertiser support for local origination programming, viewing it as *both* a service to the community and an additional revenue source. The cable company controls the content and the scheduling of L.O. programming.

Public access programming, on the other hand, is produced by volunteers and is noncommercial. In some situations, the operator provides equipment and a studio, schedules equipment and studio use, and facilitates productions. In other cases, the cable operator provides nothing more than channel time; access producers must produce their programs elsewhere and present a cassette ready for cablecast. (Special access channels are set aside in some communities for special groups, municipal, educational, arts, health, etc.) Some localities have independent access corporations (funded wholly or in part by their cable operators) to administer access activities independently.

Community programming, another term often used by cable companies, can have a number of different meanings. Because of this potential ambiguity, it is essential to understand what "community programming" can mean. The term often appears in franchise bids, usually functioning as an umbrella term under which local origination and public and other kinds of access are subsumed. The term, may, however, mean something entirely different. It may refer to a *method* of local program production. There are cable companies that do not allow public access to their systems; instead, they offer "community programming," which is exactly like local origination programming as defined above, except that community volunteers are used in some positions where local origination operations used paid staff. Volunteers may serve as producers, camera operators, writers, and even composers.

Because this term, "community programming," can function both as an umbrella term and as a name for a particular production method, it is important for members of the community to know exactly what a cable company means when it refers to "community programming." If the term appears in a franchise bid, the cable company should be asked to clarify whether the term is being used as an umbrella and, if it is, to include all budget, staffing, facilities, and programming procedure details for both access and L.O. in the bid.

If the term refers to the method of producing, the company must also be required to provide access, with its own support, equipment, studio, and budget lines.

Sometimes, cable companies will present "community programming" as a "new and innovative" type of local programming in an attempt to persuade a community that public access is not necessary. Such arguments can be very persuasive.

Dan Leahy, a former access producer who now serves as a programming consultant to Rollins Cablevision (an MSO), is an effective proponent of "community television," the Rollins version of "community programming." He explains, "I can't tell you how many loops I've been through on the issue of access, from fighting with cable operators to give me access, to promoting the concept of an independent nonprofit corporation that takes away some of the cable operator's responsibility, siphoning off a piece of the action to support it, and finding some other monies to put together an independent operation.

"I have concluded that trying to get a cable operator to act in the public interest is like trying to get a horse to do ballet. I know there are some who do it; I've seen the Lipizzaner Stallions." But requiring a cable operator to make access available is, according to Leahy, requiring a businessman to "engage in a series of operations that have nothing to do with the rate of return — requiring him to perform mandated activities that are in the public interest," but not in his own interest.

Leahy is right when he says it is difficult to get operators to cooperate in this area. But some do, and many cities in the recent wave of franchising have created franchise agreements that require meaningful access as well as meaningful access support from cable operators.

The operators are understandably reluctant to engage in an activity that robs them of control over what they feel is rightfully theirs — their channels — and costs them money but does not bring them immediate measurable revenue in return. The issues are control and income. Few cable operators are able to perceive an access operation as one that will attract viewers, although, in fact, there are good reasons to believe that access programming does.

Leahy contends, and many observers would agree, that ten years of mandated access has produced little, a "pea and shell game where the object for the operator is to *promise* things that appear to be wonderful, grand, and benevolent, but in fact offer things that leave him with no responsibility to deliver. So, it's really easy for an operator to promise a community two channels of arts programming when he doesn't have to

> "...trying to get a cable operator to act in the public interest is like trying to get a horse to do ballet."
> —Dan Leahy

do the programming himself."

Indeed, this is one franchising area in which cable operators have "definitely muddied the waters to their own advantage," commented one former cable company franchiser. Use of the ambiguous term, "community programming" was linked with an unwritten rule that said, "never use words like 'will.' We were told to say 'Such and such *could* be used for so and so,' not 'Such and such *will* be used...' or say 'The company *may* provide,' not 'The company *will* provide.' That way it sounds like you are promising, but they can't hold you to it."

What Leahy proposes in place of the "pea and shell" game of access is community programming which he says will be supported by the cable operators because it makes business sense. All operators know, Leahy says, that community programming attracts subscribers and gives them a reason to continue to subscribe. That is the first justification. But community programming can also create a new revenue stream, by giving the local merchant, producer, and service-provider a video advertising outlet that is reasonable in cost and by providing a facility that can create a low-cost video ad.

According to Leahy, this new kind of community programming involves the marriage of the community advertiser and the community event via video. The local bank, he suggests, will want to position itself as the opening and closing note surrounding the local chamber orchestra concert. Why? Because both the audience and the performers are potential depositors for the local bank.

Which Kind of Local Programming Is Right for the Arts?

Community television, as described by Leahy, may indeed be the path that cable television takes as it is threatened increasingly by competition from other telecommunications delivery systems. Community programming may be the ultimate pattern that develops for local cable productions. Some of the programming Leahy has developed as a community programmer has been very good indeed. But for those communities which are offered any kind of access provision, it is essential to understand the differences between Leahy's community television concept and the access concept, as well as the implications of these differences for the arts.

The crucial difference between access programming and community programming (and L.O. as well) is in who controls the content. As mentioned earlier, with access programming, the access producer controls the content entirely; the cable company has no control over what appears on the access channels (with the possible exception of obscenity considerations).

The second essential difference is the absence of commercial sponsorship on access channels. In order for the arts to gain access to a community programming channel, they must be able to attract the local advertising dollar. If your "product" is controversial or appeals only to an extremely narrow audience which is not a natural target for any local advertiser, your program will not be seen on community television. This contrasts with the typical access situation where, if you have made a program or want to make one, the fact that the program exists and you request time for it assures that it will be run on the television system. (It must be noted, however, that the fact that it runs does not assure that anyone will watch it.)

Sometimes both community programming and local origination look very attractive to arts organizations, because the cable operator does almost all the work—produces the program, schedules it, and sells the advertising time. And, there are situations in which this is the only way a particular production can be put on local television. In fact, there have been some very profitable collaborations between local origination staffs of cable companies and arts organizations. The arts organizations must, however, be aware that in most L.O. and community programming situations, the cable operator who produced the piece retains all rights to it; anyone involved in such a production must be prepared to compromise on that point.

Another telling point often made about community and local origination programming is that it tends to be much higher in quality (more like broadcast television) than is the case with much access programming. Often, L.O. programmers have better equipment and are more skilled in its use. Access centers are staffed for the most part by trained volunteers and usually have less sophisticated equipment, whereas with L.O. programming, which is supposed to make money, the operator may devote a great deal more time, effort, and money to that part of his system.

Finally, it must be reiterated that the cable company controls the content of L.O. and community programming. If the company objects to your program idea, it will not be produced.

George Stoney spoke recently of the situation in Canada, where nearly all community programming is created by the cable operators (but does not have to be commercially sponsored). Their programming, he said, was all "rather weak; it lacked bite and challenge. It was polite and sincere, but *noblesse oblige* does not produce strong programming." And he added, "when people are constantly *done for*, they don't

**Citizens have a right
to demand clarity in
the agreements that
their elected officials
enter into on their
behalf.**

develop strong muscles." Stoney wants artists to make their
own video. "When they depend on other people to do it, they
are losing control and also losing a way of learning a great
deal more about their own art. That is why I keep pushing for
the artists to join in the community access movement and for
them eventually to get their own equipment."

Sometimes using an access channel is the only way to tell a
story. At other times, the more generous resources of a local
origination studio production will serve a project best. In
each case there are trade-offs, and using the local cable television system requires learning enough to weigh the pros and
cons of all available approaches.

No matter what the local programming option, the potential to produce television at the local level will never materialize unless the cable system is designed to include these
features, and their presence is sanctioned by a well-written
franchise agreement. Arts representatives interested in being
able to use their local television system must be careful from
the very beginning to understand what competing cable companies are offering. Informed citizens have a right to demand
clarity in the agreements that their elected officials enter into
on their behalf. "Community programming" without
separate provision for public access is not acceptable.

Franchising: What Should the Arts Ask For?

The whole point of becoming involved in franchising is to
convince competing cable companies to include features
which will benefit the arts in their system designs. But what
will actually benefit the arts? The answer to that question
varies from community to community. To answer it requires
research, technical knowledge of cable, imagination, and the
ability to step back from the concerns of day-to-day management to explore cable's real potential.

The people planning arts usage of a community's cable
system must also take the time to reflect upon where their
organizations are headed and where the arts will be in five to
ten years. Because it may take as long as five years from the
start of franchising to the actual *use* of the system, people
planning arts use of the system must take a long-term view.
For example, consideration should be given to an institutional network (separate from the subscriber network) to link
all arts organizations via cable and to permit two-way electronic communication. If interactive systems become commonplace, new services will be available through cable, and
these could be put to use by enterprising arts organizations. A
shop-at-home service could provide a remote box office.

Banking via cable on an institutional link could prove both cost- and energy-efficient for small arts organizations. Climate control and security devices via an institutional link could also prove to be cost-effective investments. These future developments of current technology should all be considered, because provision must be made during franchising for the arts community to participate in such services. Such provisions, in fact, need to include free or reduced-cost modems and modulators so that organizations can actually hook into institutional networks.

The first question that will be raised when the arts community begins to talk about cable television is: "Should we have our own channel for arts programming?" The answer will vary from community to community. In New Orleans, the arts-cable coalition has no doubt that a separate arts channel will be a boon. On the other hand, in Santa Barbara, which was recently re-franchised, the arts-cable activists concluded that the community could not support an arts-dedicated channel. Every locale is different, and each arts community must conduct its own, careful research.

Establishing Access

Establishing access early in the franchising process is the overriding issue. Once independent access is established, questions of arts channels can be decided. It is even possible for a community to "grow into" an arts-dedicated channel within an access structure if there is sufficient demand. But the first priority must be to establish access channels and an independent access management organization with operating funds coming from a variety of sources, such as the city, the cable company, private donors, foundations, and corporations.

There are good reasons for establishing an independent management organization for access and developing a broad base of financial support. Access run by the cable company will inevitably be the first activity of a cable operation to be cut, or cut back, in times of financial distress.

Chuck Sherwood, former executive producer of Manhattan's Channel L Working Group (programmers for New York City's municipal access channel) and an NFLCP board member, encourages groups to work for the provision of independent access management in the franchise contract, so that the access-using community is protected from access withdrawal by fiat. "If the company controls it, they can just pull it whenever they feel like it. At least, if it is written into the franchise contract as an independent operation, it cannot

> . . . people planning arts use of the system must take a long-term view.

Sherwood sees access as "the last great hope for a common experience in a community."

be removed simply by whim. There is always a procedure for changing the contract. But then at least the community has a chance to testify as to why it is important that access be preserved."

Sherwood sees access as "the last great hope for a common experience in a community." Setting up an independent access management organization is the only way that the community's interests will be served. "The cable operator has a disincentive to serve the community because access competes with the program and pay services, and the operator is in business to generate as much revenue as possible from those services."

During the short history of access, there have been several different ways of administering it. In many locations, the cable company runs access; in some communities, there are independent access centers or access management groups which run a studio facility and establish guidelines for operation. In other communities, the city or a community organization (the library, for example) runs the channel(s). And recently, as part of new urban systems, city-run and independently administered access management corporations are being created to implement access.

Cooperation between an arts coalition, other special interest groups, and the rest of the community is essential to the formation of an independent access organization. The NFLCP's *Cable Television Franchising Primer* suggests that "as early in the franchising process as possible, it is a good idea to formally petition the city government to designate by resolution a group as the official public access or community programming organization." This should be done as soon as possible to assure that there will be provision in the request for proposals for an independently run access organization. If this is part of the cable television plan from the beginning, it will be much less likely to be scuttled later on by the successful bidder.

According to the NFLCP, the access organization should be nonprofit and open to all citizens, on a nondiscriminatory basis, for a nominal membership fee. The corporation should "have an operating structure that allows for free and open decision-making and broad representation of community interests." It should "exist solely for the public benefit and aim to educate and train the public to use the public access channel(s)."

In New Orleans, the Community Access Corporation was established at the very beginning in the original ordinance governing the franchise process. The corporation was not activated until after the franchise was awarded, but it was

understood at all times that an independent access corporation would administer the access channels. In Minneapolis,
too, a portion of the system's channel capacity will be set aside
by the cable company to be run by a nonprofit access organization to be set up by the city. The Boston Community Access
and Programming Foundation is yet another example of an
access corporation that runs independently of the cable company. Each of these corporations is in the developmental stage
as this book is being written. Already well established is the
Madison (Wisconsin) Community Access Center (MCAC),
which Jennifer Stearns describes in *A Short Course in Cable* as a
model access center. Financial support for MCAC has come
from the city, the cable company, membership fees, and
private donations.

For an access center to work successfully, it must have a fully
equipped origination studio. Equipment need not be elaborate, but must be durable and portable. There must be a paid
staff to administer the center, run training sessions, and provide professional help where necessary. There must be a
source of operating funds, and many advise that the funds
should come from a variety of sources so that the center is not
dependent on one pipeline. *CTIC Cablebooks, Volume I: The
Community Medium*, contains helpful descriptions of commun-
ity-run access management models. The National Federation
of Local Cable Programmers and the Office of Communications of the United Church of Christ (See listings of service
organizations in the appendix) are both sources of information on what will work in your community and for referrals to
models that function in circumstances similar to yours.

Should There Be An Arts Access Channel?

Once commitment to an independent access management
corporation has been made, the arts community can turn to
the question of whether there should be a special, arts-
dedicated access channel. To understand some of the reasons
for and against arts access channels, it is necessary to look at
the relationship between arts and access in general.

In situations where all access users share facilities and
channels, relationships between arts groups, artists, and access users are not always smooth. It would be misleading to
imply that they are. Some aspects of the spirit of access are at
odds with some elements involved in the process of making
art. By its nature and operation, access is a democratic activity, dedicated to treating all on a nondiscriminatory and first-
come, first-served basis. Art, while not always elitist, is not
necessarily a democratic endeavor.

Although some may continue to argue that no one watches access, it is not true.

The democratic spirit of public access embodied in the first-come, first-served scheduling policy of many access channels sometimes can cause artists to balk. Artists care about the atmosphere in which their work is shown, and some do not want their work presented between the local psychic's half hour and the pre-teen trivia call-in show.

George Stoney acknowledges that "the first-come, first-served philosophy is not working well in some places." He points out places where modifications of that policy have helped. In Knoxville, Tennessee, programming is scheduled in blocks of related subject matter, and in Aspen, Colorado, Stoney says, particular nights of the week have been dedicated to particular themes—one night is women's night and another is high school night.

New York City's Manhattan Cable has another way of modifying the strict first-come, first-served policy. Manhattan Cable has two public access channels, one of which is devoted to continuing access series where producers can maintain a regular time slot for a year or longer, enabling them to build a core of regular viewers.

There is no question that regularity is an aid in audience building. And, since "nobody's watching" is a charge often leveled at public access channels, this should not be ignored. It is demonstrably untrue, however, that nobody watches access. In fact, many access shows use a format that includes telephone interaction with viewers. Arts producers in Manhattan talk of hearing from viewers by telephone and by mail.

In 1977, Michigan State University's Department of Telecommunications received a grant from the East Lansing Cable Communications Commission to survey the local cable subscribers and access users. In 1977 and 1979, viewership of the access channels and of particular access programs was documented. MSU hockey was the winner, garnering nearly 10 percent of the cable system viewers in an average week. More than 15 percent of the viewers watched MSU hockey monthly. The access channel's news was not far behind those figures. Furthermore, the studies were able to record growth in audience size for some of the East Lansing access shows. Although some may continue to argue that no one watches access, it is not true.

Poor program quality is another troublesome access issue. Access equipment is usually less elaborate than local origination studio equipment, and this may affect the program quality. Will Loew-Blosser, access coordinator for the Anoka County Workshop in Fridley, Minnesota, expresses another concern related to the quality of programming: "I worry about the arts organization's use of cable," he said. "The needs of the arts are really different from those of the ordinary person who walks in to make a program. We are set up to make it as easy as possible for the average person to produce a program. In this particular studio, you can book an hour and a half of studio time a week, just about what is needed to make a talk show. But the needs and standards of an arts group are much higher than the standards of a local community group making a program. Arts people are acutely aware of many phases of production that the person off the street isn't concerned with. Every slight camera angle, lighting changes, backgrounds. Artists spend a lot of time on the atmosphere of a performance. In our hour and a half of studio time, you can't do that. Now, if there's enough money to go around, great. Take care of both needs — the arts and the regular folks. But balancing is hard."

Loew-Blosser's difficulty in balancing the needs of an arts organization from Minneapolis against those of "the average person off the street" may not be felt so strongly at other access operations, but for Loew-Blosser, it is a genuine issue.

Not only is the quality of surrounding programming on access channels sometimes troublesome for arts producers, the conditions in access studios themselves may not be conducive to the production of satisfactory (to arts organizations and artists) programming. This raises the question of whether the arts should have not only a separate channel, but also separate access studio space. Or, with arts channels, will the guidelines for use of access studios have to be changed? Such accomodations will most likely be worked out in the various municipalities as the access centers are created, and the demand and needs of the communities become clearer. One thing is clear, however: the relationship between the artist and access is not a simple matter.

George Stoney, when questioned about whether a separate access channel might help artists to deal with some of these issues, replied, "In some places, a separate arts channel will perhaps work. One might very well work in New York City. There is certainly enough product and enough people who are interested. I don't think everybody would be happy. I mean, who's arts? A retrospective of Dresden from the Met or work from The Kitchen? I can just see the sparks fly!"

Probably the most important point in favor of an arts access channel is that it establishes a presence. Viewers know where to turn for arts programming because it is always on the same channel, making it easier to publicize your organization's programs. Further, so-called vertical programming of a channel encourages viewers to stay with the channel

> ...the relationship between the artist and access is not a simple matter.

There are strong reasons for *not* requesting an arts channel.

even after the show they originally tuned in is over, which may not be the case if an arts program is followed by, say, a local hospital program.

One tendency to guard against in the dicussion of a possible arts channel is the "separatist" point of view: "We do not want to be on the same channel with X." George Stoney has observed that, "In a number of different places, the arts community, or those who *call* themselves the arts community, want to separate themselves from the grubby little people who want to do the Methodist choir. It's a class thing." For an arts access channel to succeed, it has to be a positive expression growing from the arts community, not a channel established in order to shut others out.

There are strong reasons for *not* requesting an arts channel. The primary one is that many communities cannot reasonably generate enough programming. Even with creative use of automated information on arts activities, it may be an impossible strain to create enough programming to occupy a whole channel.

Adam-Anthony Steg, a board member of the New Orleans Cultural Channel 28 (the arts channel established in New Orleans' franchise agreement with Cox Cable) admits that it will be difficult to fill the channel with locally produced programming during the first couple of years. "But," he adds, "each of the eighty or so organizations in the Cultural Cable Coalition has access to a library of pre-produced material, some of which would be pertinent to the New Orleans situation." Consequently, much of the early programming will not be locally produced. The French Consulate, for example, will make its entire library of some 600 hours of programming available at no charge, a boon to the French-speaking citizens in New Orleans.

Brian Owens, currently vice president for program development with Cable America, Inc., a Canadian-owned MSO with a reputation for support of access, speaks against an arts community beginning its video activity with an arts-dedicated channel. Prior to joining Cable America, Owens served for seven years as the access manager for the Austin, Texas cable system, which over time has achieved a national reputation. This was accomplished without significant support from the cable company, so Owens speaks from his experience of developing access under difficult conditions. "My main concern is that it's hard enough to get a person to watch *one* community channel without splitting that channel into ten different channels. The viewer has to learn what access is and has to learn to watch it. Viewers will watch a channel only if they understand what the channel is. That means a lot of promotion and audience building.

"Also," he adds, "I think people are very unrealistic when they say, 'I can produce thirty hours of programming for my arts channel.' It will drain the community of both energy and money, and the quality will not be very good. There is only so much you can do and do well. For us in Los Angeles, for example, that's about one to two hours each week. We get about one to two hours of incredibly good, viewable programming that does what everybody wants. What does a dance company want? To show that it is just as good as a big dance company, to attract a wider audience, to attract students to the school. It really takes time to produce this programming."

Owens does not believe that it is easier on the viewer when the arts have one channel to themselves. He strongly recommends that all access be together on one channel until it is full, then — as he says is about to happen in Austin, Texas — the arts may be the predominant programming on a second access channel. But if you start that way, the quality will not be high enough to hold the audience, he contends, and once an audience is turned off, it will never return.

Do as George Stoney suggests, urges Owens: Access channels should be treated as national parklands; they can be preserved for future use. Finally, Owens cautions that cable operators will offer an arts-dedicated channel as part of a franchise proposal because it is easier to offer a channel than it is to offer a studio.

What the Arts Need

If the arts community should not ask for its own channel, what should it ask for? To create successful arts programming, Owens says people in the arts must have training on video equipment, and there must be a staff arts access coordinator or facilitator at the access facility to do outreach and promotion. The access equipment available should be configured to allow access producers who become more accomplished to move on to better equipment. Since location shooting is sometimes necessary, a mobile van or portable equipment should be requested. Finally, the cable company and the city should establish a fund to support arts production. It is important that the funding be provided by a city-operator partnership, says Owens, not simply by the cable company.

If the arts community decides to request its own access channel, it should do so with its eyes open. Having an access channel will be of no use to the community without a studio, equipment, staff, and funding to support the production of programming.

. . .once an audience is turned off, it will never return.

If an arts-cable coalition working on a franchising request does not have someone well versed in the technical requirements of cable production, a consultant must be hired. Anita Benda Stech was the consultant to the Minneapolis–St. Paul Cable Arts Consortium, helping members to understand how much time and effort would be required to create one half-hour show per week. "It was a revelation to them," she noted. "It helped them understand there would not be terrible competition among groups for channel space. It gave them a completely different perspective." By the same token, many of the same coalition members had not realized how easy it would be to run an information calendar which, with very little effort, brought needed publicity to local groups.

Says Stech, "I was the technician who said, "Here is what you have said you want. It will take this kind of equipment and this commitment of personnel, time, and money from you to get it.' We also planned an implementation schedule which allows the arts to grow as the cable system grows and also allows us to make some mistakes along the way — a chance to get our feet wet."

If an arts channel is to be established, everything required for running the channel and producing programming must be included in the franchising request. Otherwise, the arts community will have a channel all to itself with no resources to use it.

Unscrambling Cable Franchising

Arts organizations have exerted significant influence over the terms of some cable franchises because they have understood the franchising process and have taken an active part in it. The process is complex, involving a number of steps which can take as long as several years, and those who are best informed are in the best position to see that their interests are well served.

In fairness, it must be pointed out here that even the best informed readers, who follow all the advice offered in this chapter, might find themselves at the end of the process with a less satisfactory franchise agreement than they had expected. Not all the variables can be controlled at the community level.

The steps followed by municipalities in awarding a cable television franchise fall into a relatively predictable pattern. During the *investigative phase*, the governmental authority empowered to grant a franchise sets up a committee, or group of committees, to investigate the community's needs, hire the necessary experts for technical advice, and detail the kind of cable service the city will seek. Enabling legislation is sometimes passed during this phase to outline the franchising procedures and the rules that will govern them.

Information collected during the investigative phase is distilled into a description of the kind of cable service the community requires. That description becomes the basis for a Request for Proposals (RFP), which is the official notification that the city is seeking proposals for the establishment of a cable television system. Before the RFP is issued, community comment is (or should be) solicited. If enabling legislation (also called the ordinance) has not been passed during the investigative phase, that is usually done at this point. The ordinance covers not only the steps and procedures involved in the franchising process, but also the regulatory relationship between the city and the cable operator and the minimum standards which the cable system must meet.

The RFP is a statement of exactly what kind of bids the city seeks for its franchise. It should describe precisely what the city wants in its cable system (the level of service, technical requirements, deadlines for construction, etc.), the condi-

. . .the franchising process can drag on for a number of years.

tions under which the city and the cable operator will do business, and the regulatory mechanisms that will govern their relationship. The RFP also requests certain financial and business information from the bidders.

Once the RFP has been written and enacted as part of the ordinance, bids are solicited. Generally, cable operators are allowed several months to prepare their bids. Once submitted, the bids are reviewed by the appropriate committees and technical consultants. Sometimes additional public hearings are held. Finally, the franchise is awarded, and the final franchise agreement is negotiated and signed.

In a municipality where the situation is politically complex, the franchising process can drag on for a number of years. In other cases, the investigation, planning, solicitation of bids, and awarding of a cable franchise can be completed within a year. No matter how much time is involved, local citizens should continue to make their voices heard throughout to assure that their needs are met.

The Investigative Phase

From the arts organization's point of view, the earliest part of the investigative phase is critical; initial community inquiries are made and advisory committees are formed. By the time the RFP is issued, the cable system's basic form and level of service will already have been determined.

Ordinarily the city council, the board of aldermen, or the mayor — whoever has the authority to grant the franchise — begins by appointing an advisory committee or task force to oversee the franchising process for the government. Because of cable television's complex technology, cities usually hire expert consultants as well.

The duties of an advisory committee typically include: educating its committee members and the community about cable television and the opportunities it presents; investigating community need for these services; hiring specialized technical consultants as required; presenting the information on which the RFP and cable ordinances will be based; and evaluating the proposals of competing cable companies. Sometimes these committees actually recommend which cable company the city should choose. In some cases, the committee remains intact after the franchise is granted and serves as the governmental body responsible for overseeing the performance of the chosen cable operator.

Some municipalities appoint citizen representatives to the franchising committee. In other cases, separate citizens' advisory groups are created voluntarily by the community. In order to have a voice in the franchising process, the arts community should be represented on the governmental committee or, if not there, on the citizens' advisory committee.

As is often pointed out, franchising is highly political. The only way to assure the creation of an adequate cable franchise is to develop active citizen participation in the process. Jennifer Stearns, in *A Short Course in Cable*, reminds her readers that despite all the publicity regarding cable television franchising, ". . . disasters are still possible. In Houston, franchising took place behind closed doors. The Houston newspapers provided little coverage of the franchising procedures. No public advisory board was appointed. When the franchise came up for a vote in 1979, it took the city council only fifteen minutes to award five franchises to five groups of local entrepreneurs, with no cable experience, who had carved up the city among themselves. None of the five franchises required public access of any kind." In other cities, however, there has been active citizen participation in every step of the process: Atlanta, Denver, Milwaukee, Minneapolis, New Orleans, and Philadelphia, to mention just a few.

As Stearns has pointed out, early franchising activities can take place without any publicity at all. If you live in a community that is not yet wired, stay in close contact with the municipal government regarding plans for franchising. Once the RFP is drafted, it is much more difficult (although not impossible) to force serious consideration of the arts community's needs.

Developing Leadership

The first priority for arts organizations is to attain representation on the decision-making committee or, if that is not possible, representation on a committee which is appointed by the decision-making body to offer advice and to review franchising proposals. In either case, the arts community must be represented by people who are knowledgeable about cable and the structure of the municipal government.

At the earliest stages of the process, the arts community's representatives should command enough respect to assure that when committee appointments are made, the city cannot afford to overlook them. This can be accomplished the same way that people in the arts conduct other advocacy campaigns: by letter-writing to city officials and newspapers, by speaking out at public meetings, and by rallying the support of interested groups.

In New Orleans, two leaders emerged: Sharon Litwin and Denise Vallon. Known in her community as a video artist,

Franchising is highly political.

Vallon understood cable technology; she also knew the ins and outs of city government. Litwin had been an executive producer for public television in New Orleans for a dozen years before becoming a journalist with the local newspaper.

Although neither Litwin nor Vallon set out to establish her credibility in anticipation of a cable television campaign, each had the required visibility and knowledge and commanded the respect of the city council and the cable operators. As Litwin explains, "Together, we could make a reasonable presentation to hard-nosed corporate people who in the past had been working with artists or community groups who had truly no understanding of the video system, which meant that the cable operators could just wipe them out, because whatever they offered sounded terrific. In our case in New Orleans, we were fortunate to have people with experience, who could set reasonable goals and who had organizational skills and artistic sensibilities as well."

Arts leadership has emerged in varying ways in different communities. In New Orleans, Vallon and Litwin led the efforts to secure special consideration for the arts, and together they organized an *ad hoc* cable committee.

In Minneapolis and St. Paul, the Twin Cities Cable Arts Consortium grew out of the recommendation of a study undertaken by the Minneapolis Institute of Arts. Further, the Institute allowed one staff member who was knowledgeable about cable to devote from 60 percent to 100 percent of her time monitoring the franchising process for the arts community.

In Milwaukee, the managers of the United Performing Arts Fund (UPAF) took several months to develop their positions on cable television. Just as Milwaukee was issuing the ordinance to govern the franchising process, William Murphy of UPAF and Patricia Tully of the Milwaukee Art Museum called the first citywide meeting of artists, resulting in the formation of the Milwaukee Arts-Culture Cable Review Committee.

The Greater Philadelphia Cultural Alliance (GPCA), a service organization for Philadelphia arts organizations, was urged into action in the cable franchising process by several of its board members and the Philadelphia Museum of Art, which loaned two staff members to the GPCA for the early part of the mobilization effort.

Arts communities involved in franchising follow a relatively standard pattern of activity. First, a communitywide group is formed to address the city's needs, and the arts seek representation there. Then the arts organizations form their own cable coalition to educate and poll the arts community and to create a cable plan. Often one or two people emerge as leaders, and almost invariably these leaders spend a significant amount of time in cable-related activities. One such time-consuming activity is helping the arts representatives to reach a consensus on their position.

Reaching a Consensus

Not only must arts organizations learn about cable television—how it works and what it can offer to the arts—they must also come to a general agreement on what position they will take during franchising. As diverse as most artistic communities are, reaching a concensus is not always easy. But unless the arts speak with one voice, there is little chance of being heard.

In Philadelphia, Carol Veit says her organization "has traditionally been very careful not to say we represent organizations if we really do not." As a result, the city's cultural community (led by GPCA) followed what has become a classic pattern for helping diverse interests reach a consensus.

"The RFP was just being finished when we became involved in cable," explained Veit. "We felt that at the time we didn't know enough to know what to ask for in the RFP. We needed to get a large, diverse cultural community 'on board' before we began to speak for our constituents in Philadelphia."

So the GPCA did not participate in the preparation of the RFP; instead, GPCA staff members who were working on the cable project plunged into learning about cable as fast as they could. At the same time, they planned a seminar geared to the specific needs of local arts organizations. The seminar covered the basics of cable technology and production, and the franchising process as it would occur in Philadelphia. One session demonstrated how video, as a medium, could be effectively incorporated into current programs. At the well-attended seminar, two questionnaires designed to pinpoint potential arts uses for cable were distributed. The first questionnaire was filled out by participants *en mass*, so that questions regarding terminology and meaning could be answered on the spot. The second questionnaire consisted of more in-depth questions and was designed to be discussed with staff members and later returned to the GPCA office.

The results from these two questionnaires and conversations with cable experts formed a basis for the position paper drafted by the GPCA and subsequently revised after review by a representative committee from the cultural community. The questionnaires and the review procedures were built into the process to assure that the end result—which was cir-

"Cable franchising is a black hole into which all your energy disappears."

culated to the bidding cable companies, the city council, and the mayor — would be truly representative of the community.

The franchising process in the Twin Cities of Minneapolis and St. Paul has been lengthy. It began in 1979 and, as of mid-1983, the St. Paul franchise still has not been awarded. During those years there has been much activity — both complicated and inconclusive — prompting one member of the arts community to comment that "cable franchising is a black hole into which all your energy disappears."

The Twin Cities Cable Arts Consortium was founded in July 1981 by people from the Minneapolis Institute of Arts and the Walker Art Center. With a grant from the Minneapolis Arts Commission, the consortium hired a cable franchising consultant, and a steering committee, composed of six people, held numerous meetings with local arts organizations.

"We had a core membership of around thirty arts organizations," explained Minneapolis Institute of Arts staff member, Jane Hancock. "We researched our arts position by asking our members to put on paper what uses they felt they would have for cable television, and by contacting other cities to find out what they had requested from their cable companies. We also used the ideas that came from eight months of steering committee meetings.

"We were very lucky to have the services of our consultant, Anita Benda Stech, because she really knew the field. She knew the cable companies, she knew how city hall works, and she knew the questions to ask the arts organizations." Hancock was the Cable Arts Consortium's day-to-day contact with city hall, while Stech wrote the consortium's report, which took the research results and put them into video terms.

When asked if it was hard to gain the cooperation of arts organizations, Hancock said no, but that it had been very difficult to maintain people's interest and commitment in a situation where the process seemed interminable. "As soon as we got one issue resolved, another one popped right up behind it."

Bird-dogging the franchising process involves an enormous amount of "boring organizational work — mailings, going to hearings, and so forth," Hancock explained. While maintaining the cable connection is part of her job, she has found other people's time is so filled with daily responsibilities that it is sometimes difficult to get responses. "It can be hard to convince struggling organizations of the necessity for appointing a cable liaison when there is no cable in the area." This difficulty occurs in nearly every city where the arts are trying to affect the outcome of a cable franchising award. Most often, one or two people have worked almost full-time on cable during some part of the franchising period.

William Murphy, executive director of Milwaukee's United Performing Arts Fund, when asked what advice he had for other arts communities participating in franchising, said: "The arts community, to be really successful at this, must get some of its own people to really understand the process fully. Then that has to be communicated to the whole community — either through newsletters or meetings. You cannot rely on anyone else, you just have to go out there and do it for yourself."

Another experienced observer cautions that in conducting ascertainments, arts organizations must be informed as to price tags and time requirements for potential cable activities. "It's easy to say 'Yes, we want to do an arts magazine,' if you don't know the cost in dollars, time, and energy."

In New Orleans, Litwin and Vallon made a concerted effort to gain a genuine consensus within the arts community. Litwin explained: "Denise and I did a lot of the organizing ourselves. Even though she and I have considerable credibility in this community, we understood the old adage that an expert is someone from out of town. We brought in some experts... who could make it clear that what we were after was reasonable. For our first meeting we flew in Jean Rice (a well-known cable television consultant) to address the fledgling arts coalition, city council members, mayor's office people, and representatives from the bidding cable companies."

At that first meeting, they circulated an *ascertainment* sheet to determine what the community wanted in cable television. They also asked each agency to write a formal letter indicating what it wanted to have. They then compiled all the information and bound it into a book, which was submitted to the mayor, city council members, and bidding cable companies. The resulting book was the official statement of the Cultural Cable Coalition: a formal request for a studio and equipment; a remote unit with microwave capability, and a dedicated fifty cents per subscriber (which, ultimately, they did not get) to support the operation of all this.

Vallon, who makes it clear that she is not a "big fan" of position papers, took one additional and somewhat audacious step in New Orleans. She took the RFP, as it was issued, and rewrote it to include the provisions the cultural community wanted. This move accomplished a number of things. It surprised people; it made it clear that the Cultural Cable Coalition knew what it was talking about, and it attracted the bidder's attention to the cultural community. Vallon suggests the

"It's easy to say 'Yes, we want to do an arts magazine,' if you don't know the cost in dollars, time, and energy."

Bidders were rumored to have spent as much as $2.5 million in Tucson during the franchise competition . . .

same procedure for any cultural community that wants to be taken seriously by the bidding cable operators.

To support their activities, Litwin and Vallon raised the needed funds, which included a small, but significant, grant from the Stern Fund. Even after the position "book" had been submitted, the Cultural Cable Coalition continued to work diligently at keeping its members interested and informed. At no point in the franchise process did the coalition sit back patiently waiting for results.

Keeping the Focus on the Arts

Once the position papers are written and submitted to elected officials, the responsible committees, and the bidding cable companies, there is usually a waiting period before the bids are submitted and the evaluation process begins. Judging from past experience, the cable companies tend to be very active during this time, lobbying and courting people who can affect the outcome of the franchising process. Arizona Civic Theatre Managing Director David Hawkinson characterized this time in Tucson as "our New Hampshire Primary period." Bidders were rumored to have spent as much as $2.5 million in Tucson during the franchise competition, some of which went into courting the arts establishment.

Under these circumstances, members of the arts community must walk a very thin line, maintaining visibility and strengthening their identity as a potential source of support for competing cable companies. Sharon Litwin put it this way: "Our rule was that no one was ever to state a preference for one company. Although many companies wanted us to endorse them, the tone we maintained was, 'We don't care *who* gets this contract. *This* is what we want *for the arts.*' "

There is only one hard and fast rule for the arts in franchising: *Never* publicly or privately endorse one cable company over another. Members of the arts community will be included in the elaborate wining and dining that is characteristic of franchising when they are perceived as well-connected and able to influence the outcome of the process. No matter how tempting the inducements may be to endorse one company over another, representatives of arts organizations must maintain their neutrality. No one ever knows with certainty who will win a franchise. If an arts representative endorses one company and another wins, the relationship between the franchise holder and the arts can hardly start on a positive note.

While maintaining objectivity, the arts community still must keep the needs of the arts in the news by cultivating press coverage and even writing letters to the editor. In New

Orleans, where the Cultural Cable Coalition wielded significant political power, the concerns of the coalition were given
broad exposure. Press coverage of the activities of the coalition was widespread. A major public event was planned, with
invitations sent to the bidding companies; meetings were held
with bidding companies to discuss cultural needs; and calls
were paid on city council members during the evaluation of
the bids to stress the importance of the arts.

The coalition's major public event took place at the Histor-
ic New Orleans Collection. Coalition members, interested
citizens, city councilmen, and the press were invited to hear
the competing cable companies give videotaped presentations
of their access proposals for the cultural cable channel. Public
interest in this meeting was high, in part because of the presence of Congresswoman Lindy Boggs, who introduced the
major speaker, James Bond, an Atlanta councilman who had
been active in his city's franchise.

The Cultural Cable Coalition also initiated a newsletter
which detailed its activities and explained the franchising process in clear, non-technical language. Attractively designed
and well-written, the newsletter provided the whole community with information about both the coalition's and the
city's activities. Also, during this period the coalition hosted
a regional conference of the National Federation of Local
Cable Programmers (NFLCP). All in all, public activities
and private meetings kept cultural concerns visible throughout the entire bid evaluation period, and this visibility certainly did no harm to the coalition's cause.

Evaluation of Bids

Once the bids are submitted, a complex process of evaluation begins. The evaluation process is usually conducted by
the advisory committee that drafted the RFP and by highly
specialized consultants. A provision for citizen advisory
review should be included as well. *The CTIC Cablebooks,
Volume II: A Guide for Local Policy* and the NFLCP's *Cable
Television Franchising Primer* both provide clear guidelines for
evaluating proposals. Particularly important is the NFLCP's
advice to "ignore vague promises qualified by such phrases as
'when economically feasible.' Such promises are difficult if not
impossible to enforce. . . ."

You can glean very helpful information by researching the
performance of competing operators in other cities where
they hold franchises. For example, if Live Wire Cable Company is offering bank-at-home via cable as a part of its application, find out whether the company has successfully tested

Check out the operator's delivery record on promises made to the arts constituencies of other municipalities.

that service elsewhere. If not, *your* community may be the test site, and no cable operator — no matter how well-respected — can be certain to deliver an untested service without complications.

What about access programming? Do other communities produce their own programming with the support of the cable company or in spite of its objections? Does the cable company cover all or a portion of the costs of access programming?

Is Live Wire Cable Company offering an arts channel, a studio with all the latest equipment, mobile vans, and a dedicated channel on the institutional network (as distinct from the subscriber network) to provide financial services to the arts community? If so, check out the operator's delivery record on promises made to the arts constituencies of other municipalities. Has Live Wire developed a reputation for promising the world to the arts community, courting board members who can affect the outcome of the bidding with the city council, and then dropping the arts cold once the bid is won? If so, it should be clear how to rate Live Wire on this score.

In the final stages of the evaluation process, the competing companies are compared on the basis of the position each bidder takes on each segment of the RFP. In some communities, a grid chart is used to represent each company's position on each portion of the RFP. Also, the NFLCP's *Franchising Primer* contains helpful examples of both qualitative and quantitative evaluations of bids.

Experts review the technical aspects of bids, and fiscal officers review the financial data submitted. In addition, most municipalities hold public hearings at which cable companies present their proposals followed by questions and discussion from those in attendance. Finally, the responsible municipal body or advisory committee must prepare a written report which formally presents the group's recommendations, detailing the reasons for each.

The role of the arts community during the evaluation period has varied considerably from one city to another. In Milwaukee, the Arts-Culture Coalition, after developing its position paper, was asked by the citizens' advisory comittee to review and evaluate the franchise proposals from the cultural community's perspective. The coalition formed subcommittees to examine bid positions in each of the six areas important to the area's cultural life: independently produced programs; access and local origination; rates, imported arts programs, pay-per-view and entertainment; equipment, research and development, and system design; administration of funds; and access authority.

The Cultural Cable Coalition in New Orleans hired a consultant to help evaluate competing proposals. The consultant, Don R. Smith, director of Bloomington, Indiana's Access Channel Three and then chairman of the board of the NFLCP, helped the coalition not only to evaluate bids from the cultural point of view, but also to assess the adequacy of the access provisions in various bids.

At this point in the process (if it has not been done earlier in the RFP), it is essential to assure that the words used to describe the services proposed in the bids mean the same thing to the company as they mean to the citizens. For example, a company must not be allowed to ignore a requirement for access channels by saying that this kind of programming is provided for under the rubric of "community programming." Where access is required, it needs to be clearly defined and specified. An example of what may be purposeful vagueness on the part of a cable company is currently causing worry in Cincinnati, a community with an arts-dedicated channel. An observer noted: "Warner Amex calls this arts channel a 'community communications' channel. The city and the franchise call it 'special interest access.' And we aren't sure if it is an access channel just like all other access channels or if it is a community communications channel which is controlled by the company and not the arts community." The difference is not just semantic. If the channel is an access channel, the community controls it; if it is a different kind of community channel, the cable operator controls it — content, scheduling, budget, everything.

Whatever the *official* role of the arts community in evaluation of bids, it must assess the impact of competing proposals on the arts. The results of such assessment should then be presented to those responsible for the franchise award — not as an endorsement of one company over the others, but as an estimation of which companies will be helpful to the cultural community. It *is* appropriate at this point to state clearly which companies are *not* recommended and why.

Because there is still negotiating room after the evaluation process, the arts community should not endorse any particular contender. But some companies will emerge as more responsive than others, and the arts community's evaluation can exert pressure on both the city council, as it makes its decision, and on the competing companies, as they jockey for position in the last stretch of the race. Finally, such constant monitoring means that a community's cultural needs will be less likely to disappear from consideration when the franchise is awarded and the final contract is negotiated.

> . . . it is essential to assure that the words used to describe the services proposed in the bids mean the same thing to the company as they mean to the citizens.

When there are major
issues to be resolved...
the only way to have
an impact is to keep
the issues in the public
view.

The Franchise Award

Finally, the city council (or other authorized body) an-
nounces the winner. Now the arts community can lean back,
take a deep breath, and relax—right? Not exactly.

Once a franchisee has been chosen, most cities begin a
negotiation with the finalist in which all the elements of the
final contract are made clear and specific. In some cities this
has been the most important part of the franchising process,
while in others there has been very little negotiation.

In Milwaukee, for example, one of the issues to be resolved
during post-franchising negotiations was the method for funding the independent access organization. Indeed, in July 1982,
Milwaukee's mayor reviewed, but did not sign, the city council's resolution awarding the franchise to Warner Amex Cable
Communications. His reason for not signing the resolution
was not disapproval; it was reportedly a gesture intended to
"emphasize the importance of negotiations" as a part of the
franchising process. The final agreement was given city council approval in April 1983 after ten months of negotiation.

In New York City, the negotiations that followed the fall
1981 announcement of franchise winners for the boroughs
outside Manhattan have already dragged on for a year and a
half without a final agreement. Public hearings still must be
held before final contracts can be written and signed.

There are also cities where very little negotiation took place
between the time the franchisee was announced and a contract was signed. In Tucson, Arizona, for example, Cox
Cable's bid was adopted nearly verbatim as the contract that
would govern the relationship between the city and the cable
operator, to the amazement of some observers (including the
city's citizen cable commission).

When there are major issues to be resolved and made more
specific in the post-award negotiation period, the only way to
have an impact is to keep the issues in the public view. *Village
Voice* columnist Bob Brewin has done that for New Yorkers in
his weekly column, which has provided the city's only ongoing chronicle of the contract negotiations. Although it would
be unusual for an arts coalition to be involved in final negotiations, monitoring the progress of talks between city officials
and the cable company is essential.

Such monitoring may not be easy. Often those involved in
such talks do not want any of the proceedings released to the
public, but agreements reached out of the public eye tend not
to be as responsive to public needs as they should be. Pressure
and publicity here are important tools.

Although some of the hardest work in cable television franchising takes place at the very earliest stages, those involved eventually learn that to assure a truly responsive telecommunications system requires continuing commitment. People must continue to monitor the system's operation and people have to *use* the system. That, after all, was the major reason for becoming involved in the first place.

Preparing the Arts Community to Use Cable

The worst possible outcome of arts involvement in franchising would be an arts community unprepared to use the cable system when it is turned on. If an arts consortium has negotiated a cultural channel, training and planning must begin early to prepare people to use the channel.

Once the Minneapolis franchise agreement is signed, the Twin Cities Arts Consortium will, according to its position paper, "continue to develop the format and experiment with the content of an arts calendar, an announcements service which began operating on the system servicing Eden Prairie ...in June 1982." The consortium plans to build upon the surrounding community's experiences with the arts calendar. The possibility that the Twin Cities systems will be interconnected may ultimately mean the calendar will be able to serve the entire area. A second part of the consortium's plan calls for training artists to use the production equipment as soon as the equipment and access center are made available.

In Boston, the Cultural Education Collaborative has not waited for the city's access center to become available. The collaborative has been preparing for the arrival of the cable system since the franchise agreement was signed. A project called "Cultural Access Through Cable Television" began in 1981, and has taught a number of people how to use cable television. Two staff members from each of ten cultural and ten community institutions participated. Described by Linda DiRocco of the Cultural Education Collaborative, "These participants represent ten partnerships between a cultural institution and a community agency. Each partnership is based on the needs of a community agency and uses the cultural institution to assist in meeting those needs. The project, itself, is divided into three phases."

The first phase covers cable television: technology, regulation, and access in Boston; education about cable in Boston and in other communities; and training in production-related areas. Phase two is a twenty-five week practicum in which each partnership produces three sample mini-programs and tests them at neighborhood sites, and phase three implements the programs on cable television. When this program is completed, there will be a group of trained people ready to use the access facilities provided by Boston's cable contract.

Community training is also being planned by the Cultural Cable Coalition in New Orleans. Sharon Litwin explained: "We don't want the company to open its access facilities and find out that the community is not prepared, so we will be sure that our people take advantage of the training programs that Cox offers as part of the ordinance, and we will offer our own type of training which will evolve over the next couple of months. We've asked each organization in our coalition to designate one or two people to receive hands-on experience, working in facilities that already have video equipment. Also, all of our agencies have a year to produce their first 'image' of themselves — either a slide presentation, a video, or in some cases it may be a live remote."

At the same time, the coalition in New Orleans is setting up operating procedures for the access center and monitoring the implementation of the franchise agreement between Cox Cable and the city of New Orleans. Already the cable operator has requested permission to make a "quite serious design change in the system," but, according to Litwin, the Community Access Corporation and the city council disapproved the change. It has become clear that the franchising work done by the arts coalition in New Orleans was only the beginning. Now their energies are directed toward making the agreement work for the arts and the community.

In November 1980, the city of Cincinnati granted a cable television franchise to Warner Amex Cable Communications, Inc. to build and operate a 138-channel system with 15 access channels, one of which was dedicated to the arts and humanities. Although the arts community had not played an active role in the franchising, at least one well-placed member of the city's establishment thinks that Warner Amex decided the arts community had real power to influence the franchise award, and therefore, gave the arts community "its own channel" to influence the award of the franchise.

In July 1982, with the wiring of the city proceeding, the Cincinnati Institute of Fine Arts commissioned a study regarding possible uses of cable communications by the arts on both the subscriber and the institutional networks. The study involved seventeen Cincinnati arts organizations and was conducted by Performance Resources, headed by James D. Rosenberger. A communications consultant and independent producer who knew both video and the arts community, Rosenberger did an ascertainment study, personally interviewing representatives of each arts organization in depth. Staff members answered detailed assessment questionnaires regarding their potential cable needs and resources, armed in advance with information on the potential cost of such activities in time and money.

At the same time, Rosenberger organized a conference devoted entirely to local cable programming for Cincinnati franchise area arts organizations. The conference brought cable practitioners from a number of cities and towns and showed the arts community how cable was being used in other parts of the country. Taped examples of programming from other franchises were used extensively to illustrate the variety of programming already being produced.

Each session of the conference was videotaped and edited in time for a cablecast the same evening on the local arts channel, demonstrating the practicality of using the cable system. "One of the things I hoped to show with that same-day cablecast," explains Rosenberger, "was that, for example, if the symphony has to cancel a performance because of rain, it may have a difficult time getting its message out via local broadcast radio or televsion, but the message can go right onto the arts channel and people will learn quickly to look there."

Until the conference, according to Rosenberger, many in Cincinnati thought about cable and the arts only in terms of the national cable services, Bravo and ARTS. "At the conference, they realized that they can affect the personality, the shape, and the kinds of programs and services that come from their own cable system here. Diverse groups also realized that whether they are large or small, they can cooperate, because their video communication needs are all very similar."

As a result of the cable work commissioned by the Institute of Fine Arts (which also included a conference dealing with the institutional network), it appears that the Cincinnati arts community is beginning to find ways to use the cable television system to its own and the viewers' benefit.

> **"Diverse groups also realized that whether they are large or small, they can cooperate, because their video communication needs are all very similar."**
> **—James D. Rosenberger**

How to Talk Back to Your Television System

Nearly everything so far in the discussion about franchising has dealt with planning, negotiating, and lobbying for a future cable system in a commuity. What happens after the franchise is awarded? How do you protect your interests? What happens in communities where the franchise is five to ten years old and the community is just waking up to the potential benefits of an up-to-date system?

Enforcing Franchise Provisions

A good franchise agreement must include enforcement provisions as part of the contract. The contract which articulates the relationship between the cable operator and the city should also establish a long-term oversight body. It can be an office in city government headed by a telecommunications officer (a new breed of civil servant); it can be a committee of concerned city officials; or it can even be the city council. At the same time the official oversight body is created, a citizens' advisory committee should be established to represent the viewers and the users.

All cable ordinances should detail procedures for resolving complaints. Finally, regularly scheduled performance reviews should be built into the agreement. Each of these provisions will help to protect citizens' rights and interests and will help to establish direct, clear communication between the cable operator, individuals, and the city government.

Most franchise agreements contain construction schedules with penalties that can be assessed if deadlines are missed. Even when there is an office responsible for monitoring the cable company's performance, it is important that the company and the government know that the arts community is watching the company's performance as well. The cable company's representatives who were present during the franchising process will have left town, and a new group will have taken their place. Old relationships between the community and the cable company will no longer exist. Let the new people know you are interested. The government officials responsible for overseeing the fulfillment of contractual obliga-

"The company either fires all the original franchising staff or removes everybody so there is no evidence left — nothing except what is written in the bid."

tions will have more strength knowing that there are knowledgeable citizens in the community who will support their actions.

"You must write about what happens after the franchising is finished," insisted one anonymous observer of cable franchising. "Tell people what happens when the first team leaves and the second team comes in. The company either fires all the original franchising staff or removes everybody so there is no evidence left — nothing except what is written in the bid. And everyone on the second team knows almost nothing about cable television, or as little as possible, so that they can act stupid. They have no knowledge of earlier conversations or of the contents of the bid. This second set of company reps is supposed to create upheaval and discouragement, to take the bid and build the system with a minimum amount of interference from the community and the city council — even if it comes down to lying and fabricating right in front of you."

As the system progresses beyond construction and into provision of service, the community must watch to see that provisions are fulfilled. "Enforcement is the key," explained a former cable company franchiser. "The first thing the cable company does after the negotiations are completed and the contract is signed is to sit down and say, 'Here's a contract; let's figure out a way to beat it.' "

If the terms of the contract are not met, sanctions should be invoked. For example, if the cable operator promised in the contract to certify people to use the system's public access equipment by providing basic training for the arts community via hands-on workshops, make sure this happens on schedule. Otherwise, the community can be prevented from using the access equipment. Some companies will try to make major changes in their systems once the contract is signed by asking permission to make what they describe as "minor" changes in the original plan. "Don't let them fool you," cautions an experienced observer, "They can take what is a 'blue-sky' franchise proposal and turn it into a pile of horse manure. It's really easy. All they have to do is change a few little things here and a few things there, and the whole thing is gone, shot to hell."

Another caution comes from a different observer in another city: "People have to understand that there is a special language that is used in franchising. Phrases like 'whenever possible' and 'when appropriate' drive people insane when they are trying to enforce agreements. The regulatory side of cable is a joke unless the definitions that everybody is following are all the same definitions. Disagreements can cause *years* of delay!" Certainly not all companies take this approach, but some do, and an alert citizenry can help to prevent such an unfortunate outcome.

Community Participation

People who are active in the franchising process can continue to have an impact once the cable system is operational. Adam-Anthony Steg of New Orleans talks about the letter-writers in the Cajun community. "We have the proverbial Cajun grandmother with a poisoned pen who will write a letter at the drop of a hat, and she will follow up. We are very lucky for that, because we don't want the cable operator to be able to say, 'Well, I haven't heard anything about it,' with the assumption that the programs are not being watched and appreciated."

Steg takes the idea of cooperation one step further. "The old attitude of 'us versus them' on the part of the public just isn't going to work anymore. This cable business is a risky proposition these days. Some MSOs may fail and be gobbled up by others. It is the responsibility of the community to work hand-in-hand with the cable operator to make all these dreams everyone is talking about work. It is a tremendous challenge I hope we are all equal to."

With the kind of cooperation between the operator and the user community to which Steg refers, any breach of contract or slippage in delivery of promised services can be handled smoothly. In any case, the city council or other elected officials responsible for overseeing the cable operator's performance need to know there are people in the community who care about what they are doing and will back them up. The spirit of franchise provisions can be kept alive by continuing community support. On the other hand, if everyone goes to sleep after the cable franchise is awarded, there is no assurance that the operator will live up to the promises made. Why should he, if no one wants to take advantage of them?

Cooperation among *all* groups in the community is also essential. In Tucson, Arizona, a destructive conflict occurred between the arts community and the citizen's cable commission following the award of the cable license to Cox Cable. Although this has widely been reported as a triumph for the arts, when viewed from the point of view of those who wish to preserve and protect the rights of all members of the community (including the arts) to use the cable system, it was no triumph.

Following the formal cable license award, the arts, represented by the Tucson Commission of the Arts and Culture, asked the newly-appointed citizen's cable commission for two

**"The old attitude of 'us versus them' on the part of the public just isn't going to work anymore. This cable business is a risky proposition these days."
— Adam-Anthony Steg**

In Tucson, a destructive conflict occurred between the arts community and the citizen's cable commission. . . .

arts-dedicated channels and 20 percent of the cable license fee, both to be administered by the arts and culture commission. The new cable commission had other priorities: clarification of the status of public access and establishment of an independently-run access management corporation. Consequently, the cable commission declined to grant the arts commission's request, indicating that it would return to it later, and asked the arts commission's support in sorting out Tucson's access situation.

Instead of cooperating with the citizen's cable commission, the arts commission appealed directly to the city council and the mayor (in effect, going over the cable group's head) for two arts channels and 20 percent of the license fee. After a long lobbying campaign, which according to one close observer of the situation "activated all the Tucson power centers," the council and mayor granted the arts commission's request.

One result of the arts community's campaign was that access issues remained in limbo and the cable commission was outmaneuvered by the politically savvy arts community. Nearly the entire membership of the cable commission resigned some months later, despairing of any opportunity to have a significant regulatory impact on the cable company and its services.

The Tucson case is one where the arts won a superficial victory, but actually created problems which will become increasingly apparent down the line. The prospects for access were left unclear. Since the cable operator opposes independently administered access, it would have been to the advantage of the arts and the cable commission to be unified when facing the cable operator. As it is, the community that will be using the cable system is divided and consequently weakened in any stance it wishes to take with regard to the cable operator. One long-time access observer cites this as an example of the "divide and conquer" tactics used by some cable operators to lessen the effectiveness of the access-user community.

Finally, some people in the Tucson community at large, hearing that some $500,000 (the amount estimated to be due the arts and culture commission once revenue starts to flow) has been earmarked for the arts in Tucson thanks to the city council and the cable company, have concluded that the arts "have been taken care of" by the city council and do not need any additional funds.

The story of Tucson will probably continue to develop over a period of years, but the unfortunate legacy of this first development is a split in a community that needs to be unified. To take full advantage of access channels and to gain

maximum participation and viewership, all elements of a
community must unite in the effort. Furthermore, if a cable
operator is not supportive of access, unity among users is
essential. For those reasons alone, arts organizations are advised to avoid divisive maneuvers regarding channels and
their use.

**There are some cable
companies that seem
able to withstand
almost all attempts to
regulate them. . . .**

Pressure Points

   If there are no penalty provisions built into the franchise
agreement, if no enforcement mechanism has been established,
and if there is no cooperation between the cable operator and
the community, then the municipality will have to wait for
either a rate-increase hearing (a change in the terms of the
contract which has to be agreed upon by both parties) or the
expiration of the current franchise in order to redress its
grievances.

   When the contract is subject to review, the community has
a chance to change conditions that it regards as onerous. For
example, in Lafayette, Louisiana the cable company notified
the municipality that it wanted to add a pay service, which
would have changed the subscribers' rates. That rate-change
request became the basis for a renegotiation of the cable
operator's contract. The new contract added four access channels and made provision for an independent access organization, the Acadiana Open Channel.

   When the operator requests a rate increase or a renewal of
the franchise, the citizens' committee or group of cable activists must be prepared to present the changes that they seek.
Getting such changes through the governmental body which
regulates cable can be as complicated as the lobbying that
goes on during the franchising process. Here, too, it is essential to mobilize broad community support.

   By the same token, there are some cable companies that
seem able to withstand almost all attempts to regulate them,
even during requests for rate increases. Times Mirror Cable
TV Inc. was refused a rate increase in one of its California
franchises in the 1970s. In response to the refusal, the cable
company turned off the cable system. "By 3:00 P.M. that day,
they had their rate increase," recalls a former Times Mirror
employee.

   Currently, several communities are attempting to improve
their cable system's service after franchises had been granted
and without any requests for rate increases by the cable companies. In New Haven, Connecticut, an *ad hoc* advisory
group was formed to lobby for the establishment of an access
operation that is truly responsive to the needs of the area's

diverse population. At the time this is being written, the people of New Haven have no access to their cable system. Representatives of the arts community and the local legal services organization led the formation of a broadly representative citizens' committee. The focus for community action was the approaching deadline for all Connecticut cable operators to submit their plans for state-mandated public access channels and procedures for their use.

In Omaha, Nebraska, where Cox Cable is building a new system as this book is being written, an *ad hoc* group from the arts community was assembled after the franchise was awarded, only to discover that what looked like benefits for the arts in the franchise agreement were illusory. One such "promise," to provide a dance floor for the local ballet, was interpreted by the city attorney to mean "a dance floor that was big enough" to dance on. (Dancers, of course, thought the "dance floor" meant a *marley* floor.) Concludes Liz Rixen, a member of the *ad hoc* group: "Now we have a franchise which is such that if the contract were interpreted, we don't really have anything for the arts." The *ad hoc* group is now surveying the community to see what can be salvaged for the arts in Omaha.

Oversight takes an enormous amount of time. In some ways it is even more difficult than franchising. Once the system is built and people are getting their movies and sports, cable becomes routine. This is a good reason for the existence of a citizens' advisory body to continue to review the performance of the cable operator. Cable activists and community movers and shakers must stay informed if the system is to be responsive.

Re-franchising and Renegotiation

Only when the contractual terms of a franchise agreement are under review do citizens have an opportunity to change onerous aspects of the franchise agreement. Requests for rate increases sometimes present such opportunities, and some franchise agreements contain provisions for periodic review and renegotiation during the term of the franchise. However, the ultimate review of a cable company's performance takes place when the franchise expires and the company is considered for renewal. This is called re-franchising.

By the mid-1980s, the original franchises awarded to many older systems will be expiring. In December 1981, *Cablevision* magazine sounded an alarm for the industry: "Cable's re-franchising era is about to open in earnest. Some MSOs are teetering on the wrong end of a pyramid. Their cash machines are their vintage systems, now coming to the end of fifteen- to twenty- to thirty-year franchise agreements. Simultaneously, well-publicized major market builds have schooled city fathers everywhere in what they might demand. . . ."

Whenever review, renegotiation, or re-franchising takes place, citizens' interests are best served by the presence of a well-functioning citizens' advisory committee or *ad hoc* cable group. The presence of such a group is assumed in the following discussion.

When a franchise is reviewed or renegotiated (as opposed to undergoing a full-blown re-franchising), it is advisable to set limited objectives. For example, if the operator has not been living up to the franchise agreement, this is the time to demand compliance. When an operator requests a rate increase or other kind of renegotiation, you must know what you want, what is possible, and what you will settle for. A clearly articulated set of objectives and a realistic understanding of the position of the cable operator will go far in such situations to make the negotiations smooth.

Re-franchising, on the other hand, when properly done, should be very much like franchising. The primary differences are that the community is already wired and the cable operator will have been serving the community for ten to fifteen years, during which time the company will have developed local alliances and customer good will. Despite the presence of such community feelings, it is essential to recognize that franchises do *not* have to be renewed automatically.

Although the worst fears of the industry (that re-franchising would prove as bloody and fierce as the franchising wars) had not come true by the middle of 1983, these fears persist. The re-franchising situation since the *Cablevision* article cited earlier has been relatively calm—calmer than many had expected. In fact, Nancy Jesuale, writing on the renegotiation of cable television franchises in the *CTIC Cablebooks: A Guide for Local Policy*, observes that "localities interested in competitive bids upon the expiration of a current franchise may find little, if any, serious interest from potential competitors of the current operator. In fact, there seems to be a 'gentlemen's agreement' among MSOs not to encourage competitive bidding at the end of one company's franchise agreement."

Despite the fact that re-franchising has not yet heated up (perhaps because of the "gentlemen's agreement" Jesuale suggests exists), at least one industry analyst says that it will in the near future. Alan Kassan, financial analyst for First Manhattan, suggests that during the latter half of the eighties, we "will see the networks and newspapers in the cable business. They are going to move in, saying, 'We've got the ability and financial resources to provide the advanced technology.' At that point, re-franchising will heat up again."

. . . re-franchising, as currently practiced, can bring citizens into the information age.

However, if the cable industry has its way, federal legislation will be passed to remove the power of the local municipality to regulate cable, and franchises will be renewed nearly
automatically. Over the past several years, legislation has
been introduced to transfer regulation of the cable industry to
the federal level. The industry will certainly persist in this
quest which will, if it ever succeeds, have enormous conse-
quences for the individual citizen's ability to have an impact
on the local television system.

For communities served by the older systems that do not
approach today's state-of-the-art technology, re-franchising,
as currently practiced, can bring citizens into the information
age. But just as is the case with franchising, this will not happen without an alert and committed citizenry to prod and
support city officials throughout the process.

Communities should start preparing for re-franchising at
least two years in advance. Most cable operators begin
preparing this early; in fact, at least one MSO starts its
preparations three to four years before contract expiration.
To prepare for re-franchising, the municipality needs to re-
educate itself on what constitutes the latest, state-of-the-art
cable systems. Up-to-date community-use ascertainments
also should be conducted to determine what, after a number
of years of cable service, the community now wants and can
use. At the same time, an evaluation of the current system
should be undertaken by experts. Finally, a description of
changes and new elements that will be required for the system
must be written. This process is exactly the same as writing
an RFP, and the role of the arts community is the same as it
was during the original franchising process: All the work
within the arts community should be done *before* the RFP is
written so the arts' considerations will be included.

The National Federation of Local Cable Programmers
(NFLCP) offers the following advice to municipalities on re-
franchising:

> If the municipality feels its current operator is capable
> of providing the services and system described in the
> new RFP, it can elect to give that company an advan-
> tage in its deliberations. However, the municipality is
> under no obligation to offer this advantage. The muni-
> cipality will be in a much better bargaining position if it
> opens the bidding to all interested companies.

Fridley, Minnesota, currently served by a thirty-channel
system, is negotiating with Storer Cable for a five-year re-
franchising agreement. Often, the prime objective in re-
franchising is an upgrade of the system, but the Fridley City
Council had decided to ask only for five or six additional

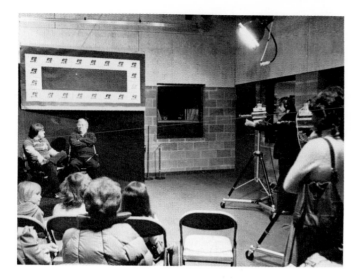

The Mayor of Fridley, Minnesota, Bill Nee, being interviewed at a 1982 open house about the re-franchising process. (Photo: Courtesy of the National Federation of Local Cable Programmers.)

channels at this point, preferring to wait and see how the 50-to 100-channel systems currently being built in other Minneapolis suburbs work out. Fridley also has an established independent access center funded in part by the town from franchise fees. (Forty percent of the franchise fee goes the the Anoka County Community Workshop, Inc. Membership fees and funds raised from diverse sources make up the rest of the workshop's operating support.) The access organization has asked for an improvement in studio facilities as part of the re-franchising and, in a spirit of cooperation not always seen in other cable systems, the independent access center and Storer's local origination program staff submitted their requests for studio facilities and equipment jointly.

When asked about arts programming, the Anoka County Community Workshop's access coordinator, Will Loew-Blosser, said, "We don't really have an identifiable arts community in Fridley." As a result, there is no separate role for the arts community in this re-franchising effort. In general, Loew-Blosser characterizes the re-franchising effort as a "pretty agreeable" one.

Both the city and county of Santa Cruz have been wired since 1954. The county's franchise expires in 1984 and the city's in 1986. Both are working cooperatively to re-franchise. In addition, five other franchises located close by will expire within the same four-year period. With "seven local franchises with a total of approximately 50,000 subscribers," consultant Thomas Karwin hopes that there will be coordination throughout the area in the attempt to get improved service, increased channel capacity, and possible interconnection of community service channels.

Communities should start preparing for re-franchising at least two years in advance.

As 1982 drew to a close, the Santa Cruz project was in the community-education stage. Residents of the Santa Cruz area were offered workshops sponsored by the city council and county board of supervisors. These workshops were set up to educate members of identifiable community groups and help them to participate in planning future cable service in the area. Special areas covered in the workshops included the arts, education, information, government, business, health, and community services. In each of these areas, there are people and groups who will benefit from participation in planning a new system that will be responsive to their needs. Following the arts workshop, the Cultural Council of Santa Cruz formed a study group which created a position paper for the arts and intends to remain active throughout the re-franchising process.

In Santa Barbara, California, where the community recently completed re-franchising, there was not as much time for the community to develop a coherent and concerted plan. Activists in Santa Barbara began to prepare in late 1980 for franchise expiration in 1982. Their first priority was to convince city authorities that renewal of Cox Cable's franchise should not be granted automatically.

Citizen's Media Watch was formed to play an advocacy role in the re-franchising. Craig Watson, then co-director of Santa Barbara Arts Services, became involved in the refranchising effort during summer 1981. He had been unaware of the impending expiration of the franchise until he learned of Media Watch's activities. "By that time, it was almost too late for us to accomplish what we wanted," he said. In addition, members of Media Watch were concerned with a multiplicity of issues, and that made it more difficult for the group to be as effective as it would have been with a narrower focus. Watson feels the community should have started sooner and concentrated on fewer issues. "Different people were interested in different issues. Some people were only interested in access and others only in protecting privacy, for example."

The new franchise in Santa Barbara is far from ideal. "Already, some of the access provisions in the new franchise agreement are breaking down," says Watson. "This community is so sophisticated, so ripe for access, and yet the provisions we have are a tiny studio and two cameras for the whole community which are supposed to be used in the field as well. Consequently, we are alreadying facing huge waiting periods for training. We failed to get provisions that would really meet the needs of the community."

"There was never any serious thought given to the creation of an arts-dedicated access channel," Watson says. "It wouldn't make sense for our community, because the built-in support would have to be *very* substantial to make it possible for our arts organizations to generate enough programming for a channel." As it is, the Santa Barbara system has three access channels: one for the public, one municipal, and one educational.

Despite the limited resources for access programming, a group of interns and volunteers, led by the Santa Barbara Arts Council, is creating an "Arts Video Team" that will produce an arts magazine for the access system. There are also some individual artists who have begun to use the access channel. Concludes Watson, "Access is still relatively new in this community, and we are suffering under the lack of space and equipment, so it is not getting the usage it could." When asked what advice he had for other arts communities in similar franchising situations, he replied, "I'd like to see more arts community involvement. In the countywide re-franchisement coming up here, I hope we are able to be more focused in our requests.

"Right now we are doing projects which we think will become success stories. We'll use them in the county re-franchising to demonstrate the arts community's level of interest and the potential pay-back to Cox from a community relations standpoint. I would like to help Cox get some acclaim outside of the community. Companies *do* have egos and it's worth stroking them. Also, the arts bring glamour and visibility to the businesses with which they have relationships. The arts have the opportunity to help a cable company look good, whether it be from entering tapes in competitions or nominating an operator for deserved awards. When Valley Cable won a Business Committee for the Arts Award, it helped them."

Clearly, Watson is suggesting that the arts establish a mutually beneficial and mutually dependent relationship with their cable operators. Re-franchising is well suited to a strategy of carrot and stick, particularly if the arts community has spent time cultivating carrots *with* the cable operator.

The Future

There can be benefits for the arts in a well-negotiated cable franchise. But there is no easy, straightforward formula that will fit every community. There are no easy answers to the question, "How can cable help the arts in my community?" In some cases, the cable system will not help at all. In other cases, with energetic people from the arts community, the cooperation of an access organization, and a good cable system, an entire new dimension for the arts may develop.

There is no easy, straightforward formula that will fit every community. There are no easy answers to the question, "How can cable help the arts in my community?"

People are amazed that televising city council hearings increases attendance at those same hearings. That is audience development for city council hearings; the same can happen for the arts. In some communities, thanks to cable television, artists, arts organizations, and viewers will learn things about the arts they never knew before. And in those communities, television itself will be better, more exciting, and more daring than it was before.

Cultivating the Wasteland

Not every artist and arts organization should have nation-wide arts exposure, but those who want to use television should have the opportunity. The arts and the artistic community have long been barred from broadcast television by the economics of the medium. Broadcast television demands a large audience; otherwise advertisers will not buy time. But not all arts programming will attract a large audience — indeed, very little does.

The economics of the nationally marketed cable services are similar to those of broadcast television, only the scale is smaller. ARTS must offer a large enough audience to interest advertisers, and Bravo must attract enough paying customers for the service to break even. Showtime will run "cultural" programming only when audience research shows that such programs attract otherwise unreachable subscribers.

The *Real* Promise of Cable for the Arts

There is more room at the local level — more room for different kinds of programming, for experimentation, for narrowly targeted programs, for diversity — more room for the arts. It is also easier to attract advertisers for cultural programming at the local level. In every town, there are advertisers eager to reach supporters of local arts institutions. In some locations, cable systems have successfully made such matches through community cultural programming.

Still, some arts programming probably will never attract advertising. In those cases, local access channels can provide an outlet by making television time available at no cost. Some cable systems (though by no means all) even provide equipment and technical assistance for people to produce access programs. In these cases, television is accessible to anyone who wants to use it.

Sue Miller Buske, executive director of the National Federation of Local Cable Programmers (NFLCP), has this advice for people in the arts: "First, get in there and use your local cable system. Second, don't get hung up on quality; think of the access channel as a blank canvas on which you will create."

Because local channels are sometimes low-budget operations, artists and arts organizations often avoid using them.
They are afraid the product will not be good enough. Good
enough for what? Network television? HBO?

The *real* promise of cable television for the arts at the local
level will be fulfilled only when artists and arts organizations
shake free of broadcast television's values and treat cable as a
"blank canvas."

George Stoney encourages artists to use video because it
shows them things about their work that they cannot discover
any other way. Local television — particularly local access
television — gives the artist a chance to show these discoveries
and the process that makes them possible.

The process by which the arts are put on, or interpreted
for, television is changing, and it is primarily the makers of
local television who are changing it. They are free to. They do
not have to please five million viewers. If they are using access
channels, they do not have to please anyone except themselves.
But this very freedom of experimentation attracts viewers.

The local level is where significant breakthroughs in arts
programming are taking place. New translation techniques
are being developed, and new forms are being invented.

Experimental work presented on local channels stretches
the ability of the viewer to relate to television. Access channels have offered exciting and unsettling new uses of the
camera in drama and dance — even for talk shows. Experimental work on local television promises to change our expectations of the medium over the next few years. Television
now is a teacher, an entertainer, a persuader, an infuriator.
Television can offer us new ways of seeing, through the work
being done in hundreds of local cable systems, access studios,
lofts, museums, galleries, and even in the streets.

Furthermore, some of the programming possible at the
local level has very practical implications for arts organizations. Dance companies, symphonies, jazz ensembles,
chamber music groups, theatres, museums, and others can
use the medium to bring subscribers, volunteers, funding
agencies, and devotées closer to the process of creating the
work. A theatre company, for example, can show its subscribers how a set was designed or how the designers worked together with the director to create a unique look and feel for a
new play. An orchestra premiering a new work in an upcoming concert can cablecast an introduction to the composition,
a layman's analysis, or an introduction to the composer; the
maestro can discuss his approach to the new work.

A cable program on new music, for example, can easily be
publicized without generating very much additional work for

the orchestra staff. All that is required is the addition of several extra lines of information to all press releases, marketing materials, and subscriber communications. A short description of the cable program, dates, times, and channels should be indicated in all these materials. One additional news release should be sent out to announce the program. Be sure all calendar editors are included in the mailing, arts as well as television. Small slips announcing the cable program and its scheduled dates and channel can be inserted in ticket envelopes by the box office staff. The program's schedule could even be printed on the outside of the ticket envelope, although more advance planning is required to do this than is often possible with access programs. In short, with minimal effort, the program's publicity can conveniently fit into the ongoing press, audience development, and marketing activities.

Creating cable television programs to introduce upcoming events deepens an audience's enjoyment of performances. The more people understand what they are seeing and hearing, the more they are able to enjoy it. Satisfied audiences are loyal; they are likely to subscribe or renew season after season. During an organization's annual fund drive, a satisfied patron is also more likely to give.

The local cable operator also benefits from this kind of programming, which generates good will toward the cable service on the part of an arts organization's subscribers and donors. The more intense the bond between the arts organization and the community, the more the cable operator will benefit from being linked to the arts organization. Arts organizations can give the cable operator a further boost by crediting the service in all program notices.

In the arts, survival has always required creative use of community resources. Not until the present has television been one of those resources. It is up to the arts to use it creatively.

Programming Costs of Local Cable

The conditions under which arts programs are created for local cable vary widely. Usually, there is a price tag attached to exploiting this resource. Using your local cable access channel as an outreach tool still requires that you produce a program, and this can cost money. There are some municipalities where it may not cost the organization anything to put on a half-hour access show: the access center or the cable company provides the facilities and the equipment, and access volunteers serve as personnel. An absolutely no-cost situation, however, is the exception rather than the rule.

Costs vary widely from one example to the next. The costs of programming discussed in this chapter vary from only the cost of the arts organization's time and labor to between $100 and $500 for a half hour of live drama to nearly $50,000. Generally speaking, however, most local programming is relatively low in cost.

Because of the wide variation in program costs from one location to the next, it is not helpful, and in some cases downright misleading, to give standard cost estimates for local programming. Arts organizations and artists wishing to become involved in local programming can find out the general cost parameters in their locales by talking with access producers (who can be located through the local access organization or cable company) and organizations which have worked cooperatively with the cable company in either local origination productions or co-productions. Some companies and access organizations also publish guides which help interested organizations to understand the range of cost and time commitment required to participate in local programming.

In some cases, a grant can help to cover production costs. In others, there may be diverse sources of funding which join in co-production. Often, however, the producers must pay out of their own pockets.

The Arts and the Cable Operators

Many communities across the country are not yet wired, but in those areas which are served by cable, locally produced arts programming is turning up more and more frequently. Some is produced by the cable companies themselves as local origination programming, some is created as community programming, and some as access programming. In Portland, Oregon, for example, Cablesystems Pacific runs the country's only local origination cultural channel. Many new franchises include provisions for arts access channels, but most of these provisions have not been implemented as this book goes to press.

The amount of cooperation from cable system operators varies. Some operators offer no public access to their system's channels. Others, required to allow access by their franchise agreements, make it as difficult as possible for community members to use the allotted channels. Some are simply indifferent. But there are a few system opertors who actively solicit public use of access channels, train people in the use of video equipment, and provide access studios, equipment, and staff to facilitate productions. The degree of cable system commitment to the arts in local origination activities is as varied as it is with access.

Some cable companies have made substantial commitments to local arts programming. One such company is Valley Cable, which serves the West San Fernando Valley area near Los Angeles. In 1981, Valley Cable gave a grant to the Los Angeles Music Center and, with the center, co-produced a pilot project for cable television. In recognition of this, Valley Cable received the 16th Annual Business in the Arts Award, co-sponsored by the Business Committee for the Arts and Forbes Magazine.

On a somewhat more modest scale, Lafayette, Louisiana's new access organization, Acadiana Open Channel, received grants from the state arts agency and a major corporation to present a weekly arts magazine program. Across the country, almost as many ways to *finance* arts programming have turned up as have different kinds of arts programming. And, as interest and resources grow, this inventiveness will undoubtedly continue.

Lack of Cable Service Hasn't Stopped the Twin Cities

Although the twin cities of Minneapolis and St. Paul still are not wired, some arts organizations have established relationships with cable systems in surrounding areas. The franchising process in the Twin Cities has been particularly frustrating: It has been going on since the early seventies and still is not concluded. To counterbalance the frustrations of franchising, Anita Benda Stech, the consultant to the Twin Cities Cable Arts Consortium, advised the arts representatives to create some programming. The Cable Arts Consortium, in turn, approached Minnesota Cablesystems-Southwest (which serves five suburban cities in the Twin Cities area) with a proposal to use the operator's Telidon (teletext) system to create an arts calendar for the Twin Cities organizations. The company agreed. Nancy Anderson, a Cablesystems-Southwest programmer, was enthusiastic: "It's cost-free and simple for the arts organizations. All they have to do is send us their calendar information and logos, and our playback operators make a Telidon slide which duplicates their logos quite well. Those logos are stored permanently, and the calendar information is changed weekly." The result is a very handsome calendar which, Stech says, any cable system would be proud to offer to its subscribers.

There were two reasons for initiating the calendar on a suburban system, according to Stech. The Cable Arts Consortium wanted to see arts-related results, and members wanted time to experiment before the Twin Cities' systems were built. "We wanted to make our mistakes early, if there

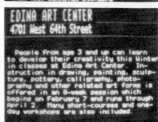

Scenes from the Twin Cities' arts calendar. (Photos: Christopher M. Anderson.)

were going to be mistakes, and we wanted to know exactly how we would do this on our own system, how much it would cost, and what was involved." The calendar has now been expanded to include arts information for the entire franchise area. "Now all the arts organizations in the [Cablesystems-] Southwest franchise area are participating," says Nancy Anderson. "The fact that the Minneapolis group started this motivated the arts groups in the community to become involved too.

"But the best result of our arts calendar," according to Anderson, "is that arts organizations which were originally nervous about cable and didn't want to get involved have now joined in. The Guthrie [Theater] was so enthusiastic about our cooperation with them on the calendar that now we're working on a number of projects for our system. When this all started, the theatre had already sold its production of *A Christmas Carol* to a national program service, but their attitude towards the *local* systems was, 'We don't want to do public access; we want to get paid for what we do.' Now, we are involved in an enormous project on their upcoming production of *Peer Gynt*."

Participation in the arts calendar project was the beginning of what some Guthrie staff members call a grassroots video movement that has emerged at the theatre during the past

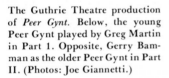

The Guthrie Theatre production of *Peer Gynt*. Below, the young Peer Gynt played by Greg Martin in Part 1. Opposite, Gerry Bamman as the older Peer Gynt in Part II. (Photos: Joe Giannetti.)

year. Tom Creamer, Guthrie literary associate, is coor-
dinating a documentary on the mounting of *Peer Gynt*, from
the production's inception to opening night. The project is being done as a public access production to be cablecast on the
Minnesota Cablesystems-Southwest's access channel. Creamer, who has had film and television training and had served
on the Twin Cities Cable Arts Consortium, was approached
by an actor, Fruud Smith, who conceived the documentary
project. Creamer took the idea to Nancy Anderson, whom he
had met as a result of the Telidon calendar project, and she
helped to start it.

Several Guthrie representatives enrolled in Cablesystems-
Southwest's video workshop to be certified to use the access
center's video equipment. Working within the access center's
rules, which limit each access producer to four uses of the
equipment per month, they established a shooting schedule.
The half-hour documentary will include interviews with
theatre staff, scenes taped in the shop during construction,
and shots of Santo Loquasto explaining his scenic designs.
Some of the research that went into the script will be
presented, along with some dramaturgical research. There
will also be rehearsal footage.

"We decided to do this tape because we thought it would be
interesting and fun," explains Creamer. "Liviu Ciulei has
encouraged it because he would like to document this produc-

tion. It will also benefit the theatre because, in addition to its being cablecast, we are going to give the Guthrie a couple of copies of the tape, and they may be used as part of the outreach program. This is not a money-making operation, however. Those of us making this tape are donating our time. Cablesystems-Southwest is providing the equipment and the editing time free as a part of their public access program."

The experience has been positive in almost every way for the Guthrie people involved. "I am continually startled and amazed at the cooperation we are getting," says Creamer. "I wonder if the company will be able to keep up this level of commitment as the demand for access grows."

Minnesota Cablesystems-Southwest is not a typical cable system operator. The company is owned by Rogers Cablesystems, a Canadian-based company which operates a number of franchises in the United States. As mentioned in an earlier chapter, another Rogers-owned company, California Cablesystems, was responsible for the modifications that made La Mirada Civic Theatre a television production facility. Rogers' systems have a track record of supporting access, and the suburban Minneapolis system is no exception. Anderson explains: "Our people are more than just available to help people learn. We literally go out and *harrass* people to learn. We *push* access. There is pressure on us all to perform, and performance means that we have access users here making tapes. This comes from corporate headquarters." This kind of corporate policy of support for access is unusual in this country.

Because of the level of cooperation with the cable operator, Creamer thinks that there will be regular Guthrie programming on public access in the future. "There are certain ready-made events which can be taped and used as educational promotion for each show. For example, before each production goes into rehearsal, there is a 'show and tell' where the director talks about the play and introduces the cast and designers to the entire Guthrie staff. That could easily be taped for public access cablecasting."

The video activity with Minnesota Cablesystems-Southwest has fostered dreams at the Guthrie, according to Creamer. "We've had 'pipe-dream sessions' where we talk about a second theatre designed to be a television studio, where we could do access programming and develop our own educational theatre programming to be sold to the local school systems and universities. Ciulei is interested; he used to be a film director and would like to do film and video again. But these are really only dreams at this point."

Although the Guthrie's involvement with Cablesystems-

Portapak training class at the Anoka County Community Workshop in Fridley, Minnesota. (Photo: Courtesy of National Federation of Local Cable Programmers.)

Southwest's public access is the most extensive, there are other arts organizations creating programs there as well. Arts access users have made dance and theatre tapes, and the local Edina Performing Arts Center produces its own promotional tapes in the cable company's facilities. "Our promotions are a little different from the usual PSA," explains Anderson, who is an arts center board member. "They are fifteen or twenty minutes long and consist of artists doing what they do. So, if an artist teaches the viola da gamba, the tape shows him playing the viola da gamba."

When cable finally comes to the Twin Cities, many of the arts organizations will have had experience with neighboring cable systems. For example, the Anoka County Community Workshop, Fridley, Minnesota's independent access organization, has allowed several Minneapolis performing groups to use the workshop's facilities to create programs for Fridley's public access channel. Also, the Minneapolis Institute of Arts has a library of tapes produced by the Institute. Jane Hancock, Audio Visal Department supervisor, says that on occasion, she allows these tapes to be cablecast on area access channels, on condition that the tapes will not be duplicated. The access coordinators love having the tapes to cablecast, she says.

Minneapolis is also home to University Community Video (UCV), which its executive director, Tom Borrup, has described as "a media arts center founded on the idea of community television. As one of the country's oldest and largest video access centers, UCV has been training video makers, making the tools of video production available, and planting seeds of community television for ten years. The video center was incorporated in 1973 in anticipation of the coming of cable to the Twin Cities. Cable didn't materialize, yet the center grew. . . ." Programs have been run to assist performing artists who want to become involved in video and, in November 1981, the center sponsored one of the first conferences held on cable and the arts.

The artists and arts organizations in Minneapolis-St. Paul are taking advantage of every opportunity to prepare themselves for the eventual completion of their own cable system. As a result, they may well become as sophisticated in their use of their cable television as they have become in the development of their other artistic resources.

Manhattan Is *Not* the Leader in Arts Use of Cable Television

The extraordinary cooperation and assistance received from Minnesota Cablesystems-Southwest for the *Peer Gynt* documentary is rare. Near the other end of the spectrum are the cable systems serving New York City's borough of Manhattan. Stubborn determination and the ability to persist in the face of daunting odds are the primary qualities shared by the access users interviewed for this book.

For franchising purposes, Manhattan is divided in half, with each half served by a different cable system. One is owned by ATC, one of the nation's largest MSOs, and the other franchise is held by Group W Cable, owned by Westinghouse. Although the franchise agreement requires that two channels be reserved for public access, there is no stipulation that operators must provide studio facilities, equipment, or training. Consequently, access users must bear the entire cost of producing the tapes used on access channels. Group W Cable has a studio which can be used to prepare access tapes (the cost at this writing is fifty dollars to produce a half-hour program). Manhattan Cable does not have access studio facilities. Access producers must produce tapes elsewhere and provide them to the company ready to cablecast. Or, access producers can rent one of several studios where live shows can be originated.

Manhattan Cable, the ATC company, has done very little local origination (L.O.) programming in the past, but says it is increasing activities now. Group W Cable produces a significant amount; a notable recent example is a very fine tape of a one-woman show, *Love to All Lorraine*. The play, based on the life of playwright Lorraine Hansberry, (*Raisin in the Sun, The*

Sign in Sidney Brustein's Window) was co-produced by Group W Cable and Woodie King.

Despite discouraging access conditions, there is a fair representation of the arts on the public access channels in New York City, including three regular dance programs, one live drama, some video art exploration, and, occasionally, music. Interestingly, there is almost no use of public access by arts institutions.

"Dance On with Billy Mahoney" is a straightforward talk show. In each segment, Mahoney hosts a half-hour conversation with one guest on a topic that relates to the lives of dancers. The format is the simplest possible, and the success of each program depends on the interest the conversation generates in the viewer. One particularly interesting recent program featured Selma Jeanne Cohen, a dance historian and aesthetician discussing her personal involvement with the evolution of dance scholarship in this country.

"Eye on Dance," also a talk show, has a somewhat more complex format. Produced weekly by ARC Video, a video studio for dance in the Washington Heights section of Manhattan, the program features a panel discussion of a dance-related issue and, when appropriate, tapes of the guests' work. Occasionally, guests illustrate points in the discussion by performing for a few moments on camera. The operations of ARC Video are funded by the state arts agency; additional funding for "Eye on Dance" has been provided in the past by the Mobil Foundation. Among the topics covered

ARC Video's weekly talk show, "Eye on Dance," during taping. (Photo: Warren Manos.)

have been "Dance on TV and Film," "Comedy and Outrage
in Dance," "Dancers' Health and Well Being," "The Business
of Dance," and "Broadway Dance." The program is marked
by intelligent discussion of issues and is geared to a dance-
involved public; the level of complexity is not diluted to make
the program accessible for everyone. The tapes, as they
accumulate, are creating a fascinating history of issues con-
fronted in the dance world of the 1970s and 1980s.

Even more adventurous visually is "Camera in the Body's
Hand," a weekly program of dance-video collaborations presented by several dancers who are committed to exploring
such collaborations. The exhaustion and frustration involved
in creating a weekly video-dance program for public access
seem to wear producers down, and since the show's debut in
1980, there has been a number of different producers. Part of
the frustration comes from the fact that the producers, themselves, must finance the programs. Although all performers,
camera operators, and crew work without pay, each weekly
segment costs from $100 to $300 for equipment rental, tape
stock, editing time, and transportation to the site of the taping, according to one of the producers, Lucy Hemmendinger.

The cable operator, Manhattan Cable, is another source of
frustration. "There is no real communication, and they just
don't seem to care," sighs Hemmendinger. At one point, she
discovered—almost by accident—that "Camera in the Body's
Hand" had been cancelled. (Manhattan access rules allow the
operator to bump a series if the time slot is requested by a new
producer, but normally, the producer of the cancelled series is
notified at least three weeks in advance and given a chance to
locate another time slot if he or she wants to continue with the
program.) Although the snafu was eventually straightened
out, and an apology for the lack of notification offered by
Manhattan Cable, Hemmendinger sees the incident as indicative of the atmosphere in which access producers work at
Manhattan Cable.

Hemmendinger is relinquishing her role as a "Camera in
the Body's Hand" producer. "It's frustrating and exhausting,
but there will be people with fresh energy to come in and pick
up behind me," she says. "The show is unique; it is an outlet
for work that deserves exposure."

Paige Ramey, a current producer of "Camera in the Body's
Hand," when asked what made her persist in creating the
show, replied, "Determination is the operative word. I like to
do it, so I'm going to continue it. This is *very, very* new work. It
reaches an audience that would never go to see these artists,
because their names are not known. But the program has interested that audience in new people; we know because of

viewer feedback. That's part of the excitement."

"Off Off Television" is a relative newcomer to the public access channels in Manhattan. Memorial Day Weekend, 1982, actress Linda Niederman was lolling on a bed watching an old Playhouse 90 drama on WNET. Why, she wondered, didn't they still do live drama on television like that? Moments later, she resolved to make it happen again herself. With no particular financial resources other than what she earned as a waitress, receptionist, and acting teacher, Niederman worked nonstop from Memorial Day to September. She assembled a group of playwrights with one-act plays, located directors and camera operators, began casting, developed a contract for the playwrights, and began the production of a weekly series of live presentations of original one-act plays on New York's public access Channel D. Originating her show at the Metro Access Studios, she has amazed everyone by producing thirteen live, half-hour dramas on a shoestring budget. Robert Patrick (author of *Kennedy's Children*), after his play *The Comeback* was produced on "Off Off Television," declared, "I haven't been so turned on by anything since I first hit the Caffe Cino."

Costs per show have varied, from a low of $65 ("That's when we brought all the props and scenery from my apartment," noted Niederman. "The show was black and white, and we paid $35 for a half-hour of air time and $30 for a tech rehearsal") to a high of $350. "It's sort of a sliding-scale operation," she explained. "People associated with the show chip in what they can. The director pays for the on-line rehearsal, if possible. Sometimes the camera operators chip in five dollars apiece." During the first thirteen-week series, Niederman was able to locate a technical director and production manager who are students at New York University's School of Film and Television; they are working without pay, but they receive school credit for their work. Niederman hopes eventually to locate a student scene designer as well.

Niederman was fortunate to have a sister, Ronnie Niederman, working at New York City's WNET on theatre projects. "Ronnie had just finished working on *Fifth of July*. She called me and said she was committted to helping with my project. I said, 'What do you mean?' 'Well, you need someone to transfer it,' she said. 'What do you mean?' I asked. 'You'll see,' she said. 'I'll be at your first production meeting.' And that meeting lasted six hours!" The most important thing Niederman learned during the first thirteen weeks of producing live theatre for television was to be "very clear with directors about budget limitations." Organizational details are important. "We have to be clear about control. It was interesting

that a lot of people who came to work with us wanted complete control — the playwright, the director — and it can't be
that way."

"Off Off Television" is not yet polished drama, but there is
evidence that the producers are beginning to develop an
understanding of the kind of drama that works well within the
limitations of the program. Quality improves from production to production. But perhaps most interesting is the fact
that a number of artists and actors have begun a television
experiment. They have found a video voice for their work.

Financing provided by the New York State Council on the
Arts made money slightly less of a problem for a recent program on Manhattan's Channel L, one of the municipal access
channels set aside in the Manhattan franchise agreement for
use by the city of New York. Channel L's users include municipal officials and nonprofit organizations. The operating
budget is provided by the two franchise holders and supplemented by foundation funding for special projects. In 1982,
the channel's producers, the Channel L Working Group, expanded the involvement of local arts organizations by presenting four arts programs.

One of these was proposed by Daniel Irvine, Lab Director
for New York City's Circle Repertory Company. Irvine
wanted to create a one-hour, live program which would
showcase one of Circle Rep's highly respected activities —
new play development. Irvine's idea was to create a behind-
the-scenes view of the taping of a scene from a play-in-
progress. Irvine recalls his conversation with the Channel L
producers: "I said to them, 'Give me a camera at the back of
the studio, I'll bring the actors out, work with them a bit to
prepare them for the taping, and tape the scene. Then the
playwright, the actors, and I will discuss the scene with the
studio audience and the television audience, using the
telephone.' They said, 'But you don't *do* that on television,'
referring to exposing the whole studio. 'That's *just* what I want
to do,' I insisted. 'Why not experiment with something that's
not usually done?' "

Irvine convinced the producers to go with his idea, and the
end result was a satisfying and interesting piece of television
that paid real dividends artistically. Irvine was able to expand
his experience as a video director, to stretch the Channel L
format beyond what the producers would normally have
offered, and to give additional exposure to the play development work done by Circle Rep.

The playwright, Bill Elverman, also found the experience
helpful in a number of ways. "People liked the dialogue, and
they understood what I was trying to do. That encouraged

Bill Elverman (above), author of *The Team*, the Circle Repertory Company production produced for cable by the Channel L Working Group, Inc. Above left, *The Team* during taping. (Photos: John D. Sandifer.)

me. Later on, I was recognized on the street. People came up to me and told me they liked my play. That was wonderful! I also experienced a very fundamental writer's transition to close-ups, and I learned from that. I always write play scripts, so it was a bonus for me to see camera angles."

As a result of the program, Irvine was able to show Elverman that his play needed more action for the stage. "The play was terrific for TV, because it is two people opening up to each other, and television makes it possible to observe that closely," he explained. "But Bill now realizes he's got to add more action to put these scenes on stage, and that realization came from the television experience."

The reaction to this modest production (it cost less than $1,000) has been enthusiastic. The Channel L producers are anxious to continue the series, and both the playwright and the director are ready to come back for more. The series will continue if further funding can be secured, and that will be an opportunity for a director, playwright, cast, and audience to learn together — something that rarely happens on television.

An Independent Access Organization Grows in Lafayette, Louisiana

Acadiana Open Channel, an independent, nonprofit access corporation, was established in Lafayette, Louisiana in 1981 as a result of a renegotiation and extension of the town's cable franchise. The independent access organization and four access channels were included in the new franchise terms on the advice of the consultant hired by the city to help with the re-negotiation process. At present, one of those channels is in use, and the Acadiana Open Channel already has "people streaming in and out of it all day," according to Tom Boozer, executive director of the Acadiana Arts Council, which operates the building that houses Open Channel and a number of local arts organizations.

Boozer describes Lafayette as a rapid-growth oil city. "Approximately 50 percent of the residents are French, very cultural people with an arts tradition that goes back over a century. The rest are wealthy, cosmopolitan people from other states." This creates a demand for arts programming that exceeds what one might expect in a city the size of Lafayette.

The programming produced by Acadiana Open Channel is notable for two reasons: In the organization's first year of operation, the programs have attracted outside arts council funding and corporate underwriting, and most of the programming is arts-oriented. "ArtsInfo," a sixty-minute, weekly magazine-format program hosted by Boozer and Lafayette Fine Arts Foundation Executive Director Michael Curry, features portions of taped local performances and interviews with artists, performers, and craftspeople. A calendar alerts viewers to upcoming arts events. The program is supported by grants from the Louisiana Division of the Arts and Aminoil, a subsidiary of R.J. Reynolds. Programming has also included: "Traditions in Cajun Music"; "The Pride of the Marsh," a barbershop quartet that specializes in Cajun humor; "A Tribute to Musical Theatre," by Chorale Acadienne; "Arts in Focus," featuring a local dance company; "Moving South, Inc." produced by the first television

Acadiana Arts Council Executive Director Tom Boozer (left) interviewing Aminoil regional manager Norm Bent (right) on the program "Arts Info." (Photo: Paul Gilmore.)

workshop as a final project; interviews with writers attending the Deep South Writers' Conference; and "Leaf by Niggle," a studio production of a short story performed "readers-theatre" style by local actors.

The operating budget of Acadiana Open Channel is provided by the city of Lafayette ($50,000) and the cable operator, Telecable ($50,000). Cliff Hall, the executive director of the organization, says he expects to be able "to raise additional funds locally for the operation of the channel. I think we are a kind of hybrid public access, fulfilling some of the functions of a local public television station because we not only receive access projects, we initiate them."

The Acadiana Open Channel has sprung up in extremely fertile soil. Efforts to diversify and broaden the sources of financial support have already borne fruit and will probably continue to do so. One workshop observer commented, "There is such enthusiasm, such creativity, such vitality. Before, there was no outlet for it. People come to the workshop and say, 'Boy, I've had this idea for years; can I do it?' "

Portland, Oregon: Home of the Nation's First Local Origination Arts Channel

An unusual franchise agreement has brought Portland, Oregon an abundance of local television channels. Cablesystems Pacific, a Rogers system, is required to program ten local origination channels, one of which is dedicated to the arts. Others provide black-oriented programming, or are dedicated to covering health, environmental issues, two channels are given over to educational services, and one channel shows the best of all the others.

Adam Haas, Cablesystems Pacific's programming director, hopes these channels will eventually pay for themselves with advertiser support. Whether or not they do, however, the system is required by its franchise to continue providing the channels. The arts community had been active in franchising, and when channels were assigned, one went to the arts. Channel coordinator Ed Geis describes the channel's operation: "Using Cablesystems [Pacific] staff, equipment, and facilities, I produce the programming. It is my decision what to cover.

"One of the first series planned by the arts channel was a cooperative effort with the Media Project, a nonprofit film distribution agency for Northwest independent filmmakers. We paid the filmmakers for their programming, and I produced a video 'wrap around' for the films that included interviews with the filmmakers on location. This gave the audience an opportunity to see the films and learn something about how and why they were made.

"The arts channel also produces a local magazine format show, "ARTSCENE," that covers music, dance, theatre, folk, and fine arts in the Portland metropolitan area. There are no critiques or reviews. The weekly show is designed to promote the arts and to get the viewer to attend performances and exhibits in the community. The entire program is shot on location. The other responsibility of the arts channel is to provide innovative programming. I call this 'art of' instead of 'art on' television. Working with local artists, I have produced programming using the electronic medium in unusual ways.

"The arts community is very receptive to our programming. I have letters from organizations saying there is a noticeable increase in attendance at concerts after a program has aired, for example. I'm not sure, however, that the arts community knows how unique it is to have a cable company sponsor local programming. Cablesystems [Pacific] has a mobile van that is capable of producing quality programming. If the artist were to rent the van to videotape a production, it would cost around $2,000 per day from a commercial production company. We provide the van and a full crew to videotape performances in the community at no cost to the performers. Cablesystems [Pacific] receives the rights to playback the tape within its system, and the performers can distribute the tape outside the system for profit, if they so choose. This has become a good way for us to get good quality programming and for performers to get a tape that can be distributed as a sample of their work or as programming on other systems."

California Cable and the Arts: New Directions

California offers some interesting models for cooperation between cable operators and artists and arts organizations. In Santa Barbara, immediately following the re-franchising process, Cox Cable agreed to donate $15,000 in staffing and equipment for a documentary project proposed by the Santa Barbara Contemporary Arts Forum and funded by the National Endowment for the Arts' Museum Program. The project, which is currently being shot by a Cox crew as a local-origination series, will result in five fifteen-minute documentaries on artists whose work has been exhibited at the Contemporary Arts Forum; the programs will be cablecast on Cox's Santa Barbara and San Diego systems. An important provision of the agreement between Cox and the Contemporary Arts Forum is that the tapes will remain the property of the arts forum, and any future profits realized through licensing will be retained by the arts forum to finance

The Santa Barbara Contemporary Arts Forum and Cox Cable Video Project on artists exhibited at the Contemporary Arts Forum. Left, artist William Wiley (third from left), and above, the Cox Cable crew during taping. (Photos: Mona Kwan.)

future video works. Featured artists include William Wiley,
Raul Guerrero, Wolfgang Stoerchle, Sylvia Salazar Simp-
son, and Paul Fairweather.

Former Santa Barbara Arts Services Co-Director Craig
Watson, who functioned as a catalyst for the project, getting
it started and staying with it through implementation, comments, "Cable company staffs aren't stable, so from the first
tape to the last, our whole crew changed totally. That makes it
difficult, but our latest crew is very enthusiastic about the project." Watson explained that doing an "arts" project is
different from doing a "broadcast" production where, for example, the television crew can bring in lights and totally
change the atmosphere while it shoots. Occasionally, the Cox
Cable crew had difficulty making the adjustment. Despite the
problems, the one tape screened before this book went to
press was outstanding. It featured the work of Raul Guerrero
and included one exceedingly memorable camera shot of a
kinetic sculpture which, by itself, made viewing the tape
worthwhile.

The kind of cooperation demonstrated in Santa Barbara
between a cable operator, an arts organization, artists, and
the National Endowment for the Arts presages new methods
of financing for local cable programming. (The Massachusetts Council on the Arts and Humanities has announced a
program called "The Cable Television Project: Partnerships
in Production" for fiscal years 1983 and 1984. The "partnerships" referred to are "between the state's cultural and cable
industries.") Further, it holds out the promise that, in the
future, other communities will benefit from the documentation of local arts institutions' activities, which residents might
otherwise not even know about.

Interestingly enough, the in-kind donation of $15,000
worth of labor and equipment for the Santa Barbara Contemporary Arts Forum project was made by a cable operator
whose public access provisions are "woefully inadequate,"
according to Watson. "There is not enough equipment, only
a tiny studio, and two cameras for studio and field use, and
already there are huge waits for training to use the equipment." When the city of Santa Barbara was re-franchised,
Cox Cable agreed to set up, equip, and staff a public access
studio. [See the preceding chapter for a fuller discussion of
the re-franchising process in Santa Barbara.] Previously,
there had been no access in Santa Barbara, although Cox did
originate some local programming. With the advent of access, Cox shifted some of the local-origination programming
to public access, and as a result, the producers of some of the
shows have decided not to continue. The difference in quality

is substantial: It is more difficult to create sets in the access studio, and the technical quality of the cablecasts is inferior. "What this means is that at this point, there is a net loss in the amount of local programming," said Watson. Still, there are a number of individuals—both theatre and other performance artists—who are using the new access studio, and the Santa Barbara Arts Council has started developing a video arts magazine, which will be staffed by a group of interns—the Arts Video Team—from Brooks Institute for Photography, the University of California, and local residents.

Just a short drive down the beach and inland, in the San Fernando Valley, public access and local origination are alive and thriving at Valley Cable, owned by Cable America, Inc., a Canadian-owned MSO with 13 franchises in the Atlanta area and three in northern Los Angeles. In Los Angeles, Valley Cable gave $50,000 to the Los Angeles Music Center for a five-part series which they co-produced with the center. Entitled "Offstage!" the series consisted of half-hour programs in which Martin Bernheimer, *Los Angeles Times* music critic, talked with Beverly Sills, David Hockney, Bella Lewitsky, Michael Tilson Thomas, and Richard Chamberlain about their lives as artists and performers. One segment was nominated for the ACE Award (the National Cable Television Association award for programming) for best cable

Bella Lewitsky being interviewed by Martin Bernheimer (bottom) on the program "Offstage!," a five-part series co-produced by Valley Cable and the Los Angeles Music Center. Below, Lewitsky about to perform. (Photos: Courtesy of Valley Cable.)

In-studio preparation for a segment of "Rasgado en Dos," a thir-
ty-minute television program resulting from a year-long project
involving major Mexican poets
and Valley Cable (top). Above,
whiteface make-up being applied
to poet Allurista for taping.
(Photos: Judith Berger, courtesy
of Valley Cable.)

entertainment programming.

The contract that Valley Cable has with the Music Center
specifies that if the programs are sold (as most believe they
will be eventually) the proceeds will be put into a revolving
cash fund for future television productions by the Music
Center. The series was created by Valley Cable's local
origination production team, which was hired by the Music
Center to produce their show.

It is only fair to point out here that Valley Cable was, at the
time, in the running for the potentially lucrative Los Angeles
franchise, and a $50,000 grant to a powerful Los Angeles in-
stitution was not likely to harm Valley Cable's chances. When
this was pointed out to the vice president for programming,
Brian Owens, he replied that the Los Angeles City Council is
a professionally staffed operation which would not likely be
influenced either by a grant or the Business Committee for
the Arts–Forbes Award to Valley Cable.

Owens contrasts the use of access in Cable America's
Atlanta franchises with that in its San Fernando Valley operations, both of which have been successful: "Atlanta is very
urban; minorities are not used to being on television. They
came in and made tremendous use of access mostly to be *on*
television. They are the media disenfranchised. The Los
Angeles people are media enfranchised: they are not interested in just being on television; they are interested in *craft-
ing* television. The quality of our programming is very high.

We do a weekly arts magazine where you'll find, for example, Martin Mull coming in to talk about his painting. Everybody working on that program will be a volunteer, and the quality of that program will be better than much of what you'll see on regular television."

Access programming usually reflects the community in which it takes place. In the same way that Lafayette, Louisiana's access programming reflects the community's French influence, about one third of Valley Cable's access is devoted to the arts. There is also much that is entertainment and personality oriented.

Valley Cable's executives and access staff had enough experience to know the major problems involved in running a successful access operation. Primary among these is audience development. Owens contends that viewers have to learn to watch access — they have to be convinced. To that end, Valley Cable has done some interesting promotions. For example, when the Sugar Ray Leonard-Thomas Hearns fight was carried nationwide as a pay-per-view offering, Valley Cable bought it and ran it on the public access channel. "Fight for the Valley," an access production wrapped around the fight, ran fifteen minutes before and fifteen minutes following the fight. "Fight for the Valley" was "a comedy take-off which was so oriented to the Valley that as soon as you turned it on, you knew it was local, with local names and local personalities. So it turned people on to the public access channel," said Owens. "We found out later that we had 70 percent of our subscribers watching that program." Access producer Rowby Goren's "Fight for the Valley" went on to win the Santa Cruz Video Festival's Best Arts Program for 1981.

In another promotion, Valley Cable ran the launch of Voyager II on access with another set of specially produced, wraparound programs. "We went out to the Pasadena Jet Propulsion Lab and interviewed the scientists that put Voyager together. We got publicity for doing that, and every time we get publicity, it helps people learn more about the access channel," explained Owens.

Festivals, the community organizer's standby, are additional promotional devices used by Valley Cable. For example, during the Jewish High Holy Days, Valley access will run nothing but Jewish programming. (Using the computer file, which stores the names and interests of all those trained to use the access equipment, the access facilitators simply call up the names of those who have expressed an interest in Jewish culture, and invite them to come in to work on the planned festival). Other festivals have centered on Thanksgiving, Christmas, the handicapped, the history of Los Angeles, and

labor. Each festival reaches a particular segment of the community that might never discover the channel otherwise.

"Down the road," says Owens, "the arts could really create
some good regional and maybe even national networks from
all the programming that is happening all over the country.
Los Angeles could have a terrific arts channel if all fifty cable
systems in the area were interconnected."

Interconnects and Regional Distribution

Brian Owens is not the only person with regional and even
national distribution on his mind. There is already discussion
of creating regional interconnects for arts programming
throughout the Northeast. A Public Arts Network has been discussed for the state of New York in a study paper entitled "Working Papers for a New York State Public Arts Network," written
by William Rushton, formerly director of research for Non-
Broadcast Television in New York City. Such a Public Arts
Network would draw from arts-oriented works created on
various access channels and package the programming for a
statewide interconnect. Although the statewide interconnect
is only a dream at this time, the proposal is an interesting one.

Adam-Anthony Steg, the audio-visual agent with the
French Cultural Services in New Orleans, approaches the
interconnect idea from a different point of view: "Regional
access channels and regional networks could help to make
indigenous culture self-nurturing. Young people grow up today feeling that their culture is not legitimate because it is not
seen on television. If it's not on TV, it doesn't exist. And that
is true not only for the French-speaking in Louisiana, but also
for the Zuni (who are trying to get their own cable system) in
New Mexico and the Scandanavians in Minnesota."

As statewide and regional interconnects begin to be more
commonplace, we will certainly see more attempts to take
some of the programming that has been created for access and
local origination and give it wider distribution. Statewide and
regional distribution via cable interconnects offer opportunities for wide exposure to artists' and institutions' productions that would not be likely candidates for national
distribution.

As programming on the newly created arts channels
increases, such programming will undoubtedly be shared
from one system to the next. New channels may find it useful
to have programming created by established channels to help
fill their time and build audiences while the local arts community produces its own programming. Small gardens may
begin to dot the wasteland.

For Future Yield

As artists and arts organizations become access producers and enter into co-production partnerships with cable operators, a few words of advice may help to prevent disappointments others have experienced.

Before involving your organization in video activity, it is wise to have a clear idea of how the video activity fits into your organization's overall aims and goals. Don't get into video just to get into video. Know what you are doing, why you are doing it, and how it relates to the rest of your organization's plans and goals. Otherwise, video and cable can become destructive, energy-consuming new toys.

Think carefully before you decide to tape one of your company's live performances—not because television exposure will reduce the audience size, it probably won't. Rather, consider it carefully because it is monumentally difficult to make a successful tape of a live performance.

If you are going to put a company performance on television, listen to the advice of Merrill Brockway: "The selection is always the most important thing—is this a particularly *good* production? A manager has to look very carefully at the group. The decision cannot be mindless. If it is a dance company, some pieces don't work on television—not even for New York City Ballet. Also, an astute company manager has to look very carefully at the works in question and say, for example, 'The dancing isn't all that good. The choreography has holes in it, and it is not a very well-mounted production.' Managers have to be honest with themselves. A lot of self-editing has to go into these decisions."

If the work passes the test, Brockway continues, "then make sure you plan an unpretentious representation of the work. And, when you get to the point of planning, call in someone who has done it before to consult with you."

"You know," Brockway added, by way of analogy, "baseball is a very tricky thing to put on television. If you want to do baseball, you should at least get someone who knows how to do baseball to outline all your problems for you." Any video production an arts group undertakes deserves the same.

Organizations must have a firm grip on the costs involved

Don't get into video just to get into video.

in putting a production on television. Extra, unplanned expenses will probably arise. Have you calculated the extra staff time that will be required to organize for the taping? Who is paying for that time? Have you considered the disruption involved for both your audience and your staff? Is your facility being used for the taping? Be prepared for a general interruption of normal activities, if only because the video equipment and the taping process is fascinating to staff members who have never seen it before. People will stop working to watch.

If you are not the producer, but are working with an independent producer or local origination staff from your local cable company, do you know your producer? Are you familiar with other productions of those producing your work? Do you feel comfortable with them? Do you communicate well?

Most importantly, talk with people who have already shepherded an arts organization or project similar to yours through the process. They can save you heartaches and substantial amounts of time.

Artists and arts organizations should never begin work on a cable production without having a full understanding of the project and its requirements. The details of any agreement must be put on paper. This becomes the contract that governs the relationship between artists or arts organizations and the cable company. This may sound simple and elementary, but it often is overlooked. The results can be unpleasant, as the following story will illustrate.

During winter 1982, an East Coast cable operator initiated a local-origination community programming series. For this operator, "community programming" involves the following elements: A community member (often a staff member from a local community service agency) is invited to serve as producer for a thirteen-week series. The community, or volunteer, producer's responsibilities are to plan the series, provide content and guests, consult with the community programming department on content, concept, and scheduling for the series, and obtain all required releases for material used and from guests scheduled to appear. The operator provides a studio, equipment and operators, and technical and professional programming assistance. The agreement between the community producer and the cable operator specifies that the operator retains ownership of the programs and the rights to use or license them in any way the company sees fit.

In this case, the series was to consist of thirteen conversations with writers, hosted by a local radio talk show personality. The volunteer producer was new to cable television and its language, but very excited about the potential benefit her series could bring to her organization. Her show host agreed to participate with the understanding (mistaken as it turned out) that he was participating in a *public access* series.

The producer, because she was new to the medium and did not realize the importance of the blanket releases required of all on-camera participants in the series, neglected to secure a clearance from her host before taping began. After the fourth week of the series, a member of the cable operator's staff requested that the host sign the required release, which would give the operator ownership of the material for worldwide use in perpetuity. The host laughed and said that he did not sign blanket releases, "assuming that we would sit down and work out something more reasonable," he recalls.

When the series was completed, the host and the volunteer producer were called to a meeting with the cable company's director of community programming. The program director told the host that he would have to sign the release or the series would not be used. The host replied that he would not sign unless the operator guaranteed that the series would be shown only in the immediate franchise area and nowhere else without the host's permission. The program director refused, explaining that company policy required a blanket release. At that point, the discussion became heated. The host held out for a limited release but the program director refused to negotiate, explaining that, in any event, the tape would never be used elsewhere because the host had no talent and the programs were not good enough. The host was angered but unmoved.

Finally, the program director threatened to erase the entire series of tapes. "At that point, I nearly slid under my chair," remembered the volunteer producer. "I felt terrible, awful—the hours and hours of work that went into this. I couldn't believe it, after *all* that work!" The host held out until he realized that the community organization he wished to help stood to lose the result of many hours of labor. At that point, he relented and signed the release. Ultimately, the series was cablecast, but only after the cable operator had lost a potential friend—who also happened to be a powerful media figure in the community.

Obviously, the producer had not fully understood what she had obligated herself to do. She had signed an agreement saying that she would provide all the necessary clearances but neglected to get an essential one—primarily because she could not imagine that anyone would object to signing the agreement. As a result, what should have been one of the first steps in lining up the host did not happen until near the end of the project. Had the producer asked for clearance before tap-

"When people start being 'producers,' they are really excited. They don't want to hear that there is a price tag on everything."

ing began, the host would have realized that the program was not public access, and he could have decided then and there whether to continue. The program director, too, could have tried to work out an acceptable compromise, instead of threatening to erase the tapes. But she did not feel that her company would find a negotiated compromise acceptable.

The cable company program director says she learned a valuable lesson from the experience: "The producer didn't understand the importance of clearances. I should have double-checked to see that they were in the file, but I didn't. Now, I try even harder to make sure that people understand what they are signing. Sometimes, when they are about to sign the producer's agreement without appearing to understand fully what they are signing, I tell them to stop and take it home with them before they sign it. They *must* realize it's a legal document."

When asked what advice she had for artists and arts organizations, she said, "When people start being 'producers,' they are really excited. They don't want to hear that there is a price tag on everything. The price tag on this is that you are going to relinquish all rights. Please understand that you are doing that. Sometimes we give the contract to our new producers to sign, after we've explained all of this to them, and they sign it as though they were signing a check for a dollar. Artists take note."

Although there are no nationally accepted standard contracts for public access, local origination, and community programming productions, not all contracts demand as much as the one in the above example. The contract between Cox Cable of Santa Barbara and the Santa Barbara Contemporary Arts Forum (governing the series mentioned in the previous chapter), for example, gives ownership of the tapes and any proceeds from future licensing to the Contemporary Arts Forum. Cox Cable, in turn, will be able to use the tapes on its community channels in both Santa Barbara and San Diego. Although some companies can be somewhat flexible in negotiating contractual agreements, others—like the East Coast cable operator—are unbending.

Another example of a contract drawn to be responsive to everyone's needs is that created for the co-production of *Love to All Lorraine*. The contract for that one-woman show, written by and featuring Elizabeth Van Dyke, contains an important provision—a share for the writer-performer in any future sale *before* the producers recoup their initial investment. "We would like to recoup our costs," explained Rick Derman, Group W's executive producer, "but our primary concern is serving our franchise area. That is why we did the

show." So the writer-performer receives 30 percent of the proceeds of any sale (before breakeven), and the balance is to be split between the producers, Group W and Woodie King.

This precedent of sharing in the proceeds of any future sale before recoupment was established by Mary Ellyn Devery in her negotiations with CBS Cable for their production of *Gertrude Stein Gertrude Stein Gertrude Stein*. Everyone agrees that it is an important contractual victory for the artists involved, because if artists have to wait until after breakeven before they can share in the proceeds of future sales, the wait can be a very long one. In many cases, the company that produces the work will never, because of its accounting procedures (which include all administrative and distribution costs), break even. Devery and Pat Carroll had agreed that they would not put *Gertrude Stein Gertrude Stein Gertrude Stein* on tape unless they could receive some continuing revenue from it. They would not go ahead with CBS Cable (or any other producer) unless the contract contained this provision. CBS Cable ultimately agreed to the provision because of its strong desire to produce the work.

Local programming is rarely a money-making situation now, but that may change in the future. Certainly, it is likely that *Love To All Lorraine* will sell, and Valley Cable has a number of programs that it expects to sell eventually as well.

Elizabeth Van Dyke in the Group W Cable/Woodie King production of *Love to All Lorraine*. (Photo: R. Derman.)

With a spirit of cooperation and the willingness to work hard at creating quality programming, the marriage of the arts and the cable operators will be a productive one.

In light of the fact that budgets for local programming are very limited, experienced producers suggest that when a cable company and an artist or arts organization are trying to forge a written agreement to make a program, both parties must decide ahead of time what their priorities are — what must be retained in an agreement and where compromise is possible. Contractual arrangements vary widely, but all agreements turn on what the parties need and how much each is willing to compromise in order to get the desired result. Both sides must work together to reach an agreement in which the essential aims of each party are preserved. A clear understanding of all the details of what you are negotiating and eventually signing is essential. Then, with a spirit of cooperation and the willingness to work hard at creating quality programming, the marriage of the arts and the cable operators will be a productive one.

Looking Ahead

Reading what was written about cable during the sixties and early seventies is both instructive and intimidating for a writer trying to predict cable's future. The fervor, optimism, and social spirit of that period had pervaded the writings on cable television. Urban planners' hopes for cable were as high-minded and public-spirited as were their hopes for urban renewal in the inner cities. In the early seventies, though, several near-bankruptcies sobered the industry, and the first cable bubble burst.

It took satellite technology to reinflate cable's bubble. As cable joined the space-age pioneers, hopes were renewed, but this time not those of the social idealists. This time it was the hopes of the TV entertainers, the television people. They raced in, ready to provide every conceivable kind of entertainment to the home via satellite-delivered networks. Narrowcasting. The cable television world became positively giddy over the prospects of unlimited channels. By 1982, nearly sixty national services were in operation or announced.

Cable system operators' hopes grew as well. Companies made outrageous promises to city officials in order to gain franchises to wire cities whose market characteristics were relatively unknown. Everywhere there was the faint smell of money; profits were just around the corner. Wall Street was sure of it. Enormous risks were taken in anticipation of future profits.

Then CBS Cable closed down. Construction of new systems in the cities proved more costly and slow than many had expected. Interest rates shot up to 20 percent and higher. By mid-1983, some of the helium had escaped the cable balloon again.

Because the cable industry is still relatively young and marked by its exceedingly fast growth, it is very difficult to predict cable's future. The future is flooded with new technologies that could, if they all succeed, make cable obsolete. Direct broadcast satellites, satellite master antenna systems, subscription TV, multiple channel multipoint distribution services. Each is a new means of transmitting a signal. Each has certain advantages over cable. And no one knows how this game is going to play out. As this book goes to press in mid-1983, some of the new technologies are still untested in the general marketplace.

Still, there is no dearth of seers—even in the face of cable's rollercoaster history and baffling new technologies. These prophets tell us that our future is that of a wired nation dotted with electronic cottages. They say we will rarely have to leave home for anything. Everything will be taken care of electronically. We will work at home, shop from home, bank from home, be educated at home. We will also receive all our entertainment at home, so they say. Some of this will come true, but how much is hard to tell.

One thing is certain: Video is playing an increasingly important role in our lives. The intersection of video terminals, coaxial cable, optical fiber, and electronically transmitted entertainment, services, and information means that homes can be part of a new network of convenience and information. Because of the versatility and multichannel capacity of its technology, a cable system can make a significant contribution to its community—but only if citizens demonstrate their interest and concern. And when their access and interests as consumers are in jeopardy, they must work to preserve them.

This is not a new issue. Former FCC commissioner Nicholas Johnson's collection of impassioned essays, *How to Talk Back to Your Television Set*, should be required reading for anyone contemplating involvement in cable television. In 1969, Johnson urged citizen participation in television, at the same time recognizing the difficulties involved.

"This is not an easy road," he wrote. "All of us lead busy lives and have other things to do. You won't win any popularity contests with broadcasters, advertisers, some public officials, and powerful local citizens. You may not be warmly received by the FCC. You may have to appeal to court. You may lose. But there is little that touches our lives as consumers more than the ever-present radio and television that fills our eyes and ears—and the minds of our children. It is clearly America's number one consumer product, our most powerful consumer product, our most powerful force for good—or evil. And it is subject to our democratic control. But only if we know our rights and are prepared to fight for them. Besides, whoever said democracy—or consumer sovereignty—was going to be easy?"

Although this was written about broadcast radio and television, it applies equally to other delivery technologies such as cable.

Keeping in mind Johnson's call to action and vigilance, does cable bring genuine promise to the arts in this country?

Will cable make a difference? Yes, in a number of different ways.

Because this is a video age, the arts will be using video more and more. Conventional broadcast television has already had tremendous impact on the arts nationwide. It has expanded awareness of the arts; it has made them more popular and more widely available. Indeed, PBS's arts programming blazed the trail for each of the cultural cable services, a fact they all acknowledge.

In the early eighties, the formation of three national cable services offering culture caused a flowering of the arts on television, and even now, with only two of the three services surviving, that flowering continues.

The explosion of arts programming introduced some artists and institutions to completely new audiences, bringing both financial benefit and increased national recognition. Those are real, tangible benefits that should not be underestimated. But those benefits will be reserved, for the most part, for the "name" artists and major institutions. National cable services will give additional exposure only to nationally recognized artists and arts organizations. But local cable systems will offer opportunities to many others.

Judging from the information collected to write this book, there is increasing involvement between the arts and cable systems in many U.S. communities. Such involvement will become even more common as new systems are built, and as new partnerships mature, there will be benefits for all involved—the cable systems, the arts, and the viewers.

The kind of activity taking place varies from one location to the next. Arts use of local cable television can be purely functional or it can be creative. Some wish only to take advantage of cable's additional marketing and public relations opportunities. Others—primarily artists—are using cable to explore their art on television and to create new forms of television.

The possibilities for local cable-arts cooperation are practically limitless. Local arts organizations can present entertaining, interesting educational and promotional programming. The creation of works can be explored, the restoration of paintings or textiles, for example, can be illustrated. So can the making of musical instruments. Behind-the-scenes introductions to community institutions can bring a faceless institution alive for viewers.

On an even more pragmatic, business level (if a cable system has a well-planned institutional network and the arts are able to participate), the use of the institutional network can bring arts organizations' administrative functions into the electronic information age. Interactive networks will eventually transport the box office into the home of the ticket-buyer.

Well beyond business applications and marketing help, cable can be used to extend the reach of art itself. Video artists' works can be introduced via the local cable system. Performers and presenting organizations will undoubtedly tape examples of their work for presentations on cable systems, thereby whetting a viewing audience's appetite for live performances.

Finally, with new arts channels coming on line and increased awareness of local cable, it seems certain that artists will increase their creative use of television. As cable television time and facilities become more available to artists, they will take the medium in new directions. We can be sure of this, because artists develop new art forms. Artists experiment. That is part of what making art is all about. And the viewer will be the beneficiary.

Local cable systems are the seed beds for tomorrow's television artists. New forms will be developed in local systems where costs are lower and risks are fewer. The best will eventually be seen at the national level. But the development work will take place locally.

Can cable put the vision back in TV? Yes. It can, it does, and it will continue to do so as long as artists continue to have access to cable systems.

A Guide to Understanding Television Deals

Copyright Fundamentals

> by R. Bruce Rich

Elements of Deal Making

> by Richard J. Lorber

The Concepts of Negotiation

> by Timothy J. DeBaets

R. Bruce Rich is a partner in the New York City law firm of Weil, Gotshal & Manges. His practice includes counseling and litigation on behalf of a wide variety of clients in the areas of antitrust, copyright, trademark, and the First Amendment.

Richard J. Lorber is the president of Fox/Lorber Associates, Inc., a company which distributes television programming to a broad spectrum of markets both domestically and abroad. Among arts organizations, Fox/Lorber works as a distributor with The Kitchen, The Merce Cunningham Foundation, The Minneapolis Children's Theater, and has consulted for The State University of New York "Arts-on-Television Project."

Timothy J. DeBaets is an attorney specializing in entertainment and general business law, and is of counsel to the New York City law firm of Stults & Marshall. He is a member of the Entertainment and Sport Law Committee of the Association of the Bar of New York City, and has lectured and participated in panel discussions on various aspects of entertainment law.

This section has been specially developed for this book by Volunteer Lawyers for the Arts.

Copyright Fundamentals

by R. Bruce Rich[1]

The artist or arts group planning to engage in cable production must not only become familiar with the business aspects of such a venture, but be mindful of important legal considerations as well. One such important area of law is copyright, which provides a means for protecting the works of creative artists against appropriation by others.

The relevance of copyright law to the production of cable programs is severalfold. A cable production likely will involve a mix of both old and new elements — be it a dramatization of Albee, an adaptation of an off-Broadway production, a medley of Gershwin tunes, or a recital of Frost poetry, to cite just a few examples. Many of these creative works will be protected by copyright, which means that they cannot be used or adapted for a cable production without obtaining the permission of the copyright owner.

Equally important, copyright provides a means for the cable producer to protect his own creative efforts against unauthorized exploitation by others — to avoid, for example, someone videotaping a cable production of *Hamlet* without permission and selling copies to theatrical distributors, television networks or syndicators, or directly to the viewing public.

With the foregoing considerations in mind, this chapter provides a basic guide to the Copyright Law for those involved in creative productions — whether dramatic, musical, talk-or magazine-format, or other — which may be shown on cable television. Among the topics discussed are: the nature and purpose of copyright protection; how this protection is obtained and preserved; and how one avoids infringing the copyrights held by others. It is important to note that these discussions are necessarily general in nature; a lawyer should be consulted for specific advice regarding the application of these basic principles to specific productions.

---

1. The author gratefully acknowledges the assistance of Lois Eisenstein, Esq. and Debra Kass Orenstein, Esq. in the preparation of this chapter.

What Is Copyright?

Copyright law is based upon the premise that, in order to encourage original works of authorship, the government should grant to the creators of literary, artistic, dramatic, motion picture, musical, and choreographic works (among others) the exclusive right to exploit their works for a limited period of time (the creator's life plus fifty years for works created after January 1, 1978, according to the Copyright Act of 1976).[2] Generally speaking, this exclusive right to exploit comprises the right to reproduce, adapt, publicly distribute, perform, and display the protected work. The copyright owner is, of course, free to permit others to exercise one or more of these rights, through assignment or the issuance of appropriate licenses (at whatever fee arrangements are determined to be appropriate), but the use of a copyrighted work without the owner's permission will generally be regarded as an infringement of copyright, and the unauthorized user is subject to the penalties prescribed by the Copyright Law.

What Is Copyrightable?

To qualify for copyright protection, a creative work must be original and fixed in tangible form (e.g., a writing, photograph, film, tape, or another form of notation). The requirement of originality is not rigid or difficult to meet. To qualify as original, a work must be the independent creation of the author and must reflect some minimum degree of originality and creativity. Virtually any cable production involving even the most minimal degree of creative effort will likely qualify for copyright protection — at least to the extent of enabling the producer to prevent others from exploiting that which is original to the production. Thus, for a production which employs pre-existing copyrighted works (e.g., involves a dramatization of a copyrighted play), the producer's original contribution — the dramatization — will be entitled to copyright protection.

At the same time, there are certain types of efforts which have been regarded as involving such minimal levels of originality and creativity that they do not qualify for copyright protection. Examples include the mere creations of names,

---

2. Under the previous copyright law, governing works created prior to 1978, copyright protection extended for an initial period of twenty-eight years, renewable for an additional twenty-eight-year period. The 1976 Copyright Act amends those time periods in the case of works still in their initial or renewal copyright term to entitle the copyright owner to a total of seventy-five years of copyright protection. For works created anonymously, pseudonymously, and "works for hire" (explained under the heading "Obtaining Rights from Others"), the term is seventy-five years from first publication or one hundred years from creation, whichever expires first.

titles, and slogans (which may, however, qualify for trade-mark protection), familiar symbols or designs, and simple listings of ingredients or contents.

How to Obtain Copyright Protection

Assuring copyright protection involves several simple steps. First, while not required by the Copyright Act to do so, the cable producer would be well-advised to register his work with the Copyright Office. By doing so, he will be entitled to take legal action against an infringer and have a broader range of remedies than would be otherwise available.

The procedure for registering a copyright is simple and in-expensive, and copyright applications include easy-to-follow instructions. Forms correspond to the type of work being reg-istered and are available from the Copyright Office (Register of Copyrights, Library of Congress, Washington, DC 20559). For published or unpublished pictorial, graphic, or sculptural works, Form VA should be used; for published or unpublished audiovisual works, Form PA should be used. Any copyrighted work can be registered, regardless of whether it has been pub-lished, by submitting an application along with a registration fee of ten dollars and the prescribed number of copies of the work. The application form states the number of file copies required, which depends upon the type of work and whether it is published or unpublished.[3]

Second, once the work has been "published," appropriate copyright notice should be affixed to all copies of the work. Under the Copyright Act, the concept of publication has a special meaning. A work is "published" when copies or phonorecords of the work are distributed to the public "by sale or other transfer of ownership, or by rental, lease or lending." While, under the law's definition, public performance or display of a work does not in itself constitute publication, licensing the work to a producer of cable television program-ming, or leasing time on a local cable channel to transmit the work, would amount to publication.

When a work is published within the meaning of the Copyright Law, the owner assumes several obligations in order to protect his rights properly. First, and most impor-

---

3. If the copyrighted work is an unpublished motion picture or a published or unpub-lished motion picture soundtrack, a potentially money-saving provision is made for submitting an "alternate deposit" of the work. For an unpublished motion picture, the alternate deposit can be either an audio cassette of the soundtrack or a visual repro-duction from each ten-minute segment of the film, accompanied by a description of the motion picture setting forth specified information. For a motion picture sound-track, a transcription of the work, or a tape or record containing the entire work, accompanied by reproductions of frames of the film showing the title, soundtrack credits, and the soundtrack's copyright notice, if there is one, will suffice.

tantly, the Copyright Act requires that a copyrighted work, when published, must contain appropriate copyright notice on all publicly distributed copies. Unlike failure to register, failure to provide the required notice may result in the loss of copyright protection. If a work has been published without appropriate copyright notice, it may be possible to cure the omission, but this will depend on the particular circumstances under which the omission occured. In seeking a remedy, the copyright owner should consult with a lawyer.

The copyright notice consists of three elements: the symbol © (lower case "c" with a circle around it) or the word "Copyright," or the abbreviation "Copr."; the year of first publication; and the copyright owner's name (or otherwise generally known designation of the owner, such as a pen name), or an abbreviation by which the name can be recognized. When the copy is a phonorecord or a sound recording, the three elements of notice are the same, except that the symbol ℗ (lower case "p" with a circle around it) should be used in place of the symbol ©.

The requirements for placement of the copyright notice vary, depending upon the form in which the work is submitted. In the case of phonorecords, the notice should appear either on the surface of the record itself or on the label or container. For "visually perceptible" copies of a work—such as scripts, videotapes, or filmscripts—the law requires that copyright notice be affixed to the copies "in such a manner and location as to give reasonable notice of the claim of copyright." Although the copyright law does not require that the notice be placed on any specific part of the work—only that the placement be "reasonable"—the Register of Copyrights has issued guidelines for suggested placement with respect to particular types of works. Thus, for example, for motion pictures, videotapes, or other audiovisual works, notice can appear near the title or the cast and credit listings, immediately following the work's beginning, or immediately preceding its ending.

Finally, the copyright law requires that two copies of a published work be deposited with the Copyright Office within three months after publication. While failure to make such a deposit will not invalidate the copyright, it may subject the copyright owner to a fine. The deposit requirement is satisfied if copies of a published work have already been submitted for the purposes of copyright registration. If the work was submitted for registration in *unpublished* form, the deposit requirement will *not* have been satisfied if the work is subsequently published, and two additional copies of the published work should be deposited within three months of publication.

Obtaining Rights from Others

Whether a cable television production constitutes a totally original, "made-for-cable" production or draws upon pre-existing works in whole or in part, the talents—and corresponding copyright rights—of a number of separate artists are likely to be involved. As noted earlier, among those entitled to copyright protection for their works are: the creators of literary and dramatic works—e.g. novelists, playwrights, historians; composers of music; choreographers; visual artists; motion picture and television producers; and record companies for their sound recordings.

If, therefore, as an ambitious cable producer, one decides to create a made-for-cable musical comedy and is depending upon the creative talents of others to write a script, create a musical score, and choreograph the production, each of those contributors potentially has a copyright interest in his or her work. The producer must be sure that the necessary copyright rights have been obtained to allow the production to go forward.

There are a number of possible ways to obtain the rights needed. One possibility is to obtain all copyright rights in the works. This may be accomplished in one of two ways. Where a work has been "specially ordered or commissioned" as a part of an audiovisual work (e.g., a script or musical score written for a cable television production), the Copyright Law will treat the contribution as a "work-for-hire" if the parties have agreed in writing to so treat the work. In such a case, the copyright is owned by the person or entity (the cable producer) commissioning the work. (Similar work-for-hire treatment is accorded to works prepared by an employee within the scope of his or her employment; the copyright in such works is owned by the employer.) Where the parties' relationship does not result in work-for-hire treatment for the creative works, the producer still can obtain ownership of the copyright by securing a written and signed agreement of the copyright interests from the creator.

Outright ownership of copyright rights by the producer—particularly in the work-for-hire setting—is not uncommon in theatrical and television productions and allows the producer to maintain exclusive and perpetual rights without need to seek permission for subsequent uses of the works. Depending upon the contractual arrangements between the producer and the creative artist (as may be affected by particular industry guild arrangements), such subsequent uses may or may not require the producer to make additional payments to the artist.

The alternative to securing outright ownership of the

copyrighted works is to obtain a license from the copyright owner covering the proposed use. This technique is commonly used in connection with pre-existing copyrighted works.

Many cable productions will rely, at least in part, upon pre-existing copyrighted works. Those works may be adapted into what the Copyright Law calls "derivative works" (e.g., a dramatization of a novel, a cable television production of an off-Broadway play, a musical arrangement of a song), or simply used in their original form. In either case, however, permission of the copyright owner must be obtained prior to use or adaptation of the works. To the extent the derivative work involves originality and creative effort, the creator of the new work will also be entitled to copyright protection for the new aspects of the work.

Licensing arrangements are a common means of allowing pre-existing copyrighted works to be used for limited purposes. For example, as a cable producer you can request permission of a songwriter (or his agent) to use one or more of that songwriter's compositions specifically in connection with a particular production. Typically, the license will provide for payment of an agreed-upon fee which will cover a specified use, time period, and medium. The license may provide for a one-time, lump-sum payment to the copyright owner or may call for payments keyed to various uses, e.g., first and subsequent showings on cable television. If the rights obtained are exclusive in nature, (i.e., the user is the sole person to whom the particular rights may be licensed), the license should be written and signed. Non-exclusive rights may be granted orally. As a matter of sound practice, the producer should make certain that the copyright owner does in fact own the specific rights he is licensing and that those rights have not been granted exclusively to another party.

How to Investigate the Copyright Status of the Work

In planning to use an existing work as a part of a new production, the cable television producer must first determine whether the permission of the copyright owner must be secured. In dealing with plays, novels, poetry, sheet music, and other published works, published copies of such works should contain a copyright notice, which will include the date of publication. From this information one should generally be able to determine whether the copyright is still in effect, based on the terms of copyright discussed previously under "What Is Copyright."

If no copyright notice appears, this may mean that there is

no copyright covering the work, but one cannot be sure; the copyright notice may have been mistakenly omitted and, since publication, the owner may have cured the omission. It should be noted that if one is misled by the absence of a copyright notice, he will not be held liable for damages for infringing acts that occur prior to his receipt of actual notice that the work has been registered with the Copyright Office. Even an innocent infringer could, however, be required by court order to relinquish the profits attributable to the infringement. And, once notice has been provided, it may be determined that production must cease or, in the case of a complicated program, that it might not be telecast or retelecast unless permission is obtained from the copyright owner.

If copies of the work are unavailable, it is still possible to find out if the work is registered through a search of the public files of the Copyright Office. (Remember, however, that failure to register a work does not necessarily deprive an owner of his copyright.) A number of libraries around the country have copies of the Copyright Office's Catalog of Copyright Entries. The catalog is divided into the different categories of copyrighted materials and lists all registrations made during particular periods of time. It is important to remember that such a search may not locate a rights assignment, whereby the owner of a copyright may have transferred his interest in the work to another party.

Rather than conducting the search oneself, a producer may wish to engage a professional copyright-search firm or an attorney who specializes in this service. Also, for a fee of ten dollars per hour, the Copyright Office will search its own records.

Once a producer has found out who owns the copyright, he can contact the owner directly in order to secure permission to use his work. In many instances, the owner will refer the producer to his attorney or agent.

In the case of proposed uses of pre-existing music, the licensing rights are typically controlled by the songwriter's music publishing company. That company might agree to license terms in a direct negotiation, or it might refer the producer to a licensing agent, such as the Harry Fox Agency, which licenses music rights for theatrical and television productions on behalf of thousands of music publishers.

It is important to recognize that the traditional practice for music licensing in the television industry has been for producers to obtain from the music publisher what is commonly known as the "sync" rights—the right to record particular compositions onto film or videotape. The sync right alone does not, however, authorize subsequent over-the-air or cable broadcasts of the music, for which an additional performance rights license is required. Traditionally, performance rights licenses for over-the-air broadcasts have been obtained directly by television broadcasters, rather than by the producers at the time they obtain sync rights licenses, through three so-called performing rights societies — ASCAP, BMI, and SESAC.

These traditional arrangements have now become the subject of a court challenge which has yet to be finally resolved as of the time of this writing. If the position of the television broadcasters in this litigation is sustained, producers of over-the-air programming will no longer be able to rely on their customers — the local television broadcasters — to obtain performance rights licenses from ASCAP, BMI, and SESAC. Instead, it would be expected that the producers would obtain not only sync rights, but performance rights as well, at the time they negotiate for their music requirements.

The ultimate outcome of the litigation is likely to have an impact on music licensing arrangements for cable television. Local cable system operators, as of this writing, do not have licensing arrangements with ASCAP, BMI, and SESAC and, therefore, a producer of programming for cable television cannot rely upon their outlets to obtain performance rights. The safest practice, therefore, would be for the producer to obtain *all* necessary music rights — including performance rights — at the time the music is acquired. To the extent that a cable production may be distributed through one of the cable network suppliers, such as HBO, the producer should check with that supplier to ascertain whether it has made any arrangements with ASCAP, BMI, or SESAC covering music performance rights. To the extent that it has, the producer may be relieved of the need to obtain those rights directly from the copyright holder.

When Copyright Permission Is Not Necessary

While the general rule is that the author's or artist's permission is needed before another may make use of his work, there are exceptions: The work may be in the "public domain," or the planned use of the work may be defended as legal "fair use" or fall within one of the exemptions set forth in the Copyright Act.

Works in the Public Domain

Works for which copyright protection has either expired or been lost are said to be in the public domain; these works can be used freely by anyone. The situation can become complicated, however, if there is an existing derivative work. Since a derivative work can be copyrighted separately from the underlying work, the fact that the underlying work has entered the public domain does not deprive the new version of copyright protection for all original material added to the source material. Conversely, the fact that a derivative work has fallen into the public domain does not automatically make using or showing it safe, as the underlying work may still be protected by copyright.

Fair Use

The "fair use" defense, developed by the courts and now codified in the 1976 Copyright Act, permits limited use of copyrighted works for the purpose of criticism, comment, news reporting, teaching, scholarship, or research. An example of what may be considered fair use is the reading of a brief passage from a Robert Frost poem as a part of a critical discussion of Frost's work. Another example might be the showing of a short scene from a Fellini film as part of a program exploring the art of filmmaking.

Since the availability of the fair use defense depends upon the facts of a particular situation, it is impossible to provide a blanket rule whereby one can be certain that his planned use of copyrighted material will be considered "fair." One should be aware, however, that many courts have taken a restrictive view of the doctrine. Before proceeding with the use of copyrighted material without permission, a lawyer should be consulted.

Statutory Exemptions

A cable program producer who plans to adapt an existing copyrighted work for use on cable may in some cases be able to use the work, without permission, under exemptions provided in the Copyright Act. Exemptions that might apply to works used on cable exist for (1) noncommercial transmissions to the blind or deaf, and (2) religious programs. To qualify for the first exemption, the program must involve the transmission of a nondramatic literary work designed specifically for blind or hearing-impaired individuals. In addition, the program must be nonprofit; there cannot be direct or indirect commercial advantage to the performers, promoters, or organizers. The exemption for religious services applies only to performances or displays of nondramatic literary or musical works in the course of religious services. However, the showing of motion pictures and other audiovisual works, even those which have a religious or philosophical theme, are not within the scope of this exemption.

Compulsory Licensing Provisions of the Copyright Act

As noted, the broadcasting of a copyright work constitutes a public performance and, as such, requires the authorization of the copyright owner. With the advent of cable, copyright owners asserted that cable re-transmissions of broadcast signals also constituted public performances and, therefore, required separate authorizations. The cable industry argued, on the other hand, that their systems functioned merely as "community antennae" and that, accordingly, the cable system acted as a receptor rather than as a performer.

The position of the cable operators was affirmed, in 1968 and again in 1974, by the United States Supreme Court, which held that the cable systems performed, in essence, a viewer function and, therefore, that such cable re-transmissions did not constitute "performances" within the meaning of the Copyright Act. These rulings were legislatively overturned in the Copyright Act of 1976, which extended copyright liability to such cable re-transmissions. However, the House Committee recognized that it would be "impractical and unduly burdensome" to require each cable system to negotiate with each copyright owner whose work is re-transmitted. In light of this, a "compulsory license" system was established whereby cable systems were automatically entitled to re-transmit programs upon compliance with certain statutory requirements and the payment of prescribed royalties to the Copyright Royalty Tribunal. The determination of royalty fees under the compulsory licensing system is based on a formula keyed to a percentage of the gross receipts of the cable company and the number of re-transmissions it makes of distant signals.

Eligible copyright owners (i.e., those whose works have been included in a re-transmission by a cable system) may file a claim with the Copyright Royalty Tribunal for a share of the fees. The Act permits claimants to agree among themselves to an appropriate division of the fees and to designate a common agent to receive payment on their behalf, without running the risk of a violation of the antitrust laws. The Tribunal is empowered to resolve disputes concerning fees to be distributed; to date, the proceedings involving the Tribunal have been marked by bitter disputes, protracted settlements, and complicated litigation.

It is important to note that the provisions of the compulsory license only apply to programs originally aired on broadcast television and then re-transmitted by cable. Programming which originates on cable, e.g., satellite-delivered cable network programming or local access shows, is not covered by these statutory provisions. Accordingly, the precautionary copyright steps outlined above must be followed with respect to such programming to avoid infringement claims.

Infringement and Remedies

If a copyrighted work is used without permission (and if such use is not a fair use or otherwise exempt from liability), the copyright owner is entitled to sue for copyright infringement. To win an infringement suit, the copyright owner must prove both his ownership and that unauthorized copying has taken place. Copying may be inferred where the unauthorized user had access to the work, and where the allegedly infringing work is substantially similar to the copyrighted work. In determining substantial similarity, courts will apply the standard of whether an ordinary person viewing both works would think that one had been copied from the other.

The copyright owner may seek an injunction — an order that the infringement cease — and impoundment of illegal copies. He may also elect to recover his actual damages plus the infringer's profits, or statutory damages. The exact amount of statutory damages is determined by the court, within a prescribed range of $250 to $10,000 for non-willful infringements; if willful infringement is proved, the maximum is increased to $50,000. In addition, criminal penalties may be imposed if willful infringement was committed for commercial advantage or private gain.

Registration has an important effect on a copyright owner's rights and remedies against an infringement. A work must be registered before an infringement suit can be filed, although a suit may concern infringements that occurred before registration. However, where registration occurs more than three months after the work was first distributed or displayed commercially, and the infringement took place before registration, a court can only award two types of relief — an injunction and actual damages — plus the infringer's profits. Unavailable to the copyright owner in cases of pre-registration infringement which occurs more than three months after first publication are the so-called "extraordinary remedies": awards of attorney's fees and statutory damages.

Conclusion

The foregoing discussion should serve to alert the aspiring

cable producer to the basic copyright concepts with which he or she should become familiar. Cognizance of one's own copyright rights, as well as the rights of others, will serve to maximize the benefits to be derived from one's work, as well as avoid unnecessary, costly, and time-consuming legal problems.

Elements of Deal Making

By Richard J. Lorber

The new golden age of television, with its cultural glamour and promises galore, is already over. Television's royal carriage of the arts turned back into the pumpkin almost before the ball began. Not only was there no prince, no slipper, no wedding, but many art groups are still paying for their ride to the ball.

Still, local, regional, and national arts groups *can* form partnerships domestically and abroad to create their own programming. But first, a question must be asked: Why should any arts group want to be involved in television at all?

There seem to be at least three good reasons: experience, exposure, and earned income. But does any *one* of these reasons really warrant the full plunge into television? The upside of the opportunity in each case palls before the downside dangers. Going into television for experience risks frustration, failure, and a costly waste of time. Going for exposure risks damage to an organization's image, an alienated audience, and jeopardized funding. Going for earned income risks financial devastation, defection of talent, and compromise of artistic integrity.

Since the downside risks are so grave, it is probably better to divert energy elsewhere unless at least two of these reasons equally reign. Certainly, earning income should never be an arts group's sole motivation for getting involved in television.

Assuming the motivations are sound, how does an arts group select and develop a project for television? Simply put, how do you get from ideal to deal? Again, the key is motivation. If exposure and experience are paramount, many opportunities exist for the arts to secure some measure of involvement in television at the local level. Where money is not the motivation, the artistic freedom and integrity of an arts goup are usually safest. Entering the business of television, however, casts a very different light. Deals can compromise ideals—but they do not have to.

Packaging the Property

The vocabulary changes when an arts group looks to television as a marketplace for their wares. Before any deal can be developed, an arts group must soberly evaluate what they have to sell in this market and how that "product" or "property" can be made marketable. Fact: Major dollars stand to be earned only in pay and broadcast television. The networks, national public broadcasting, and the two or three pay television services which finance or license anything other than feature films want stars, a "hot property," a big event — some handle with enough popular appeal to make the program promotable.

Promotability is the key to packaging and the driving force behind most deals. For commercial networks, promotability means tune-in, rating points, and ad dollars. For public broadcasting, promotability means pledges, tune-in, and assured exposure for underwriters. For pay television, promotability is satisfying subscribers with the value of something other than theatrically-promoted feature films.

With this in mind, an arts group must assemble the kind of package which will command attention in the marketplace without compromising the artistic integrity of the project or group. Packaging entails assembling talent, producer, financing, distribution, or any subset of these elements.

An arts group which is not of national or international stature can leverage its position in the marketplace by several packaging strategies. First, it can appeal to a major talent to be part of the television production, either as a performer or host. Frequently, stars have rejected direct offers from pay television and even the networks, but may be willing to associate themselves with a cultural cause. They can thereby generate revenues in support of a favorite arts group and enhance their own public image. Secondly, the cultural organization can synthesize a big event (often a pseudo-event), which may entail creating and ceremonially conferring a special award, or contriving a festival or contest. Usually, these forms of packaging require validation by an authoritative association. Such an association can also attract talent not available to the arts group alone. Third, the group can contract to perform or present some major property — a lost play by an important, perhaps recently deceased playwright, for example. Any exclusivity or newsworthiness which can be claimed for the vehicle will enhance its promotability and improve chances for successfully packaging the production. Here, the *vehicle* is the star — i.e., that lost play or never-before-seen work of choreography. The star vehicle also has the power to attract a major performing star.

Talent is perhaps the most crucial consideration in packaging, because a celebrated personality, a star, or a popular host is the most identifiable element in a production. Frequently, it offers the most effective hinge for attaining wide exposure and provides great incentive to a potential source of financing. Arts groups may be able to attract a personality who is a member of the board, or an artist's friend, thus providing the ribbon around the package which gives it promotable visibility.

Once the key element in the package, the star talent, has been secured, it becomes easier to assemble the other elements of the package — the producer, the financing, and perhaps the distribution. A good place to start is with the producer. Presuming that an arts group will want to control the production, any buyer of programming will be especially concerned that a product they intend to commit to purchase or finance will have production values consistent with other programs they produce. A programmer will raise serious questions about the commerical sophistication, experience, and ability of an arts group to deliver a professional television production. Television programmers do not come from the arts world, so they are looking for what Hollywood calls a *bankable* producer, someone who has a track record of delivering commercial-quality productions which get bought for television and appeal to viewers.

In some cases, a producer who has been working in the commercial marketplace may cross over and begin to get involved with arts organizations. For example, the fact that Robert Altman is perceived as bankable and promotable made it possible for him to sell the concept of producing two one-act plays for the ARTS cable service by a virtually unknown playwright named Frank South. Unfortunately, you can count on one hand the number of Robert Altmans that may be available to arts groups. Nevertheless, many qualified producers can be attracted to an arts organization if it has already obtained that first magic element — a star.

Roles of a Producer

The word producer is a broad and loosely used term in television which describes individuals who carry out a range of responsibilities and functions in developing a property. An executive producer is the individual, or the group, that puts the deal together and, in most cases, is responsible for financing the deal. An executive producer may do nothing more than make a few phone calls to secure the signatures of a few of the interested parties. When it comes down to recruiting stars, the executive producer may have to be the first aboard the package, acting as, if not actually being, an agent.

Apart from the executive producer, there is an important role to be played by the line producer. Specifically responsible for the conception and execution of the physical production, the line producer constructs the budget; hires the director, the talent, and the crews; supervises the production; and basically becomes the bridge between the creative and business sides of the whole endeavor. In some cases, the producer also is a capable director. More often than not, in taping a live performance, the director will be behind-the-scenes, determining the camera shots and mentally anticipating the post-production editing. With the producer, the director may maintain a very active role in shaping the physical production.

Other roles relevant to the packaging and execution of a production include the associate producer, who may function as an adjunct to the executive producer or the line producer. In cases where there are many hands putting together the deal, associate producer credits may be passed around rather freely. The major defined role of production manager is akin to the line production function. Occasionally, the production manager will assume the full role of the producer in managing the physical production.

Selecting a Producer

The first step for all arts groups, in seeking a producer and conceiving a production, is to watch television intently and diligently, studying the programs' credits as well as the programs. If programs can be taped and examined over a long period of time, arts groups will better understand the different approaches to producing and directing and be able to make better choices on their own programming path. Programs produced locally by broadcast stations and even cable access shows should be watched closely since they often achieve creditable results with limited resources.

No one should be shy; simply get on the phone and call the local broadcast station, cable system, or even the national network to track down an able and appropriate producer. (As most people are now aware, because of the discussion of the repeal of the financial interest in the syndication rules, most network programming is produced by independent producers, not staff members of the networks or broadcast organizations.)

Producers are almost always independents and are usually looking for work. The major studios in Hollywood have development departments which also can direct arts groups to producers. But in approaching Hollywood or any of the major commercial television operations, if an arts group does not already have a star on the hook, it is doubtful that a concept alone will be sufficient bait for a top-notch producer. Even brilliantly original concepts are commodities in oversupply in the creatively superheated programming business.

Furthermore, there are not many opportunities to originate programming through only the two cultural cable programming services, Bravo and ARTS. Bravo produces a very limited amount of original programming, mostly jazz and music shows done by staff members. ARTS has curtailed most original production at this point and has cut back on helping to finance productions which are in development. Foreign markets, PBS, and the commercial broadcast or pay television networks offer limited but more broadly based opportunities.

If an arts group is strong enough to make a good play for earning income in television, it should seek a commercially established producer who may also have a commitment to the arts. Occasionally, a group may be able to use its own "in house" producer if he has enough credits to be convincing to the buyer and if the package can attract financing because of the assembled talent.

An alternate route for securing a producer is to approach another important player in the television business. Distribution companies get involved in the process of packaging and may be attracted to a property which an arts group originates if it has star commitments or controls a major property. Distributors are often instrumental in putting whole packages together as executive producers. They also work with and represent many producers who know the marketplace, and they can project the marketability of individual programs.

Development and Financing

Without going into detail about a particular case, assume that an arts organization has developed an attractive package. For a major program buyer—an HBO, an ARTS, an NBC network—the question remains: How will financing for the production be obtained? Typically, a prime-time network program produced by an outside production company, such as Lorimar or Paramount, will cost an average of $600,000 to $1 million per hour. Programs produced for pay television, such as entertainment specials, variety shows, or small-scale theatrical productions (but not the scale of a *Camelot*, budgeted considerably over $1 million), will typically cost between $300,000 and $500,000 per hour, not much less than network. Original programs produced for the cultural cable services will cost in the $100,000 to $250,000 range.

Production costs should always be assessed in terms of the potential for recouping those costs and for earning income from *ancillary* or *aftermarket* sales (discussed later in this chapter).

How does an arts organization secure the money for a production? As the package is being assembled, one possible step is to approach the end users of the program. The producer or the executive producer may take the lead role in putting the deal together.

The executive producer presumably has contacts with program buyers with whom he has previously done business. Such a producer can get the "commissioning editor" of music programming at England's new Channel 4 to return a telex, can sit down at lunch with the director of programming of ARTS, or can set up a meeting with the programming decision-maker for PBS's series "American Playhouse," all of which are reasonable outlets for arts programming on television.

Quite simply, the producer will ask these program buyers whether there is a place in their budgets and schedules for a particular property. If there is any interest, the producer will immediately follow-up with a presentation of the project. This may be a two- to twenty-page proposal which will include a short description of the proposed program and background on the parties involved—the arts organization (with press clippings), the talent, the stars (with letters of intent), brief bios on key individuals, the producer with that important list of credits. (Although it may not be submitted at this time, the arts organization and the producer also will have prepared a preliminary budget, which indicates the costs to produce the program on the level they expect to achieve.)

After the proposal has been received, buyers will usually ask a series of questions: How firm is the commitment and availability of your star and producer? Can you deliver stronger or more definite letters of intent from them or their agents? If the package involves a very desirable property, what rights can you guarantee? For example, can you present letters from the estate of a deceased playwright that authorize you to exploit this property? If a program is going to involve music, archival footage, or any other elements which create special rights and clearance problems, can you give some preliminary guarantee that these elements will be available at a reasonable cost? All these questions will be used to qualify the proposal in the buyer's mind.

If an arts group gets to square one with the program buyer, many opportunities will open up. These may include development money or at least letters of intent, if not commitment, from a program buyer. In turn, such documents can be used to induce other investors to commit amounts of money to develop the property further. This may involve scripting, production of a short (ten minute) "promo" on the program or even a full-length pilot production—something to indicate what the final show or series will be like. The money also may be used simply for "pre-production," to cover the necessary expenses of securing the other elements in the package, whether it involves travel to scout a location, designing sets, rewriting the script, or making a downpayment to lock in a facility where the program is to be taped.

At this point, the program buyer may say that he likes your idea very much, but that he is either not in the position, or not inclined, to finance the production. The most difficult deals to achieve, especially for arts groups, are those which involve end-user financing of the complete production. A cable program service, network, or local broadcaster assumes its greatest risk in financing a production and may prefer to buy, that is *license*, the program after it is produced. If they love the package and the arts group cannot find the money to produce the show elsewhere, they may have to put up the financing in order to have the program produced. More often, programmers prefer to see the program finished at someone else's expense.

Of course, this can be very discouraging for an arts organization, but it is not the end of the world. Contacts, at least, have been made with the end users, or program buyers, and can be leveraged to attract outside investors, whether they are board members, friends of the star, or canny corporate heads who see earning potential for at least recoupment and a payoff in prestige "p.r."

It should be noted that investment in cultural programming has been notorious for its lack of strong payout potential. Very few cultural programs have had a track record of high earnings for their investors, and it is doubtful that anyone has become rich investing in cultural programs. However, it can be argued that quality cultural programs have a long shelf life and such classics may generate revenues at least as long as most feature films. (This is another reason why any production that is mounted should be crafted to hold up well over time.)

But apart from earning potential, production financing which comes from the corporate sector is usually structured as a contribution, rather than an investment, and serves to sup-

port public television. With the demise of CBS Cable and
The Entertainment Channel, both of which had commitments to cultural programming, public television remains
one of the last strongholds of the arts on television for the
American public. There are about three hundred public television stations around the country which are part of the
Public Broadcasting System (PBS). Most of them originate
some form of programming, and five or six of the biggest stations produce most of the programming which is seen
nationally.

Outside producers face a practically nil probability of PBS
picking up an entire production tab. Except for a special producing and financing unit like "American Playhouse," most
programs are licensed or only partially financed by PBS. PBS
may love a proposed show, but the producers will be asked to
find underwriters to support public television's desire to
finance it. Consequently, if the program is geared toward
public television, it is wise for producers to begin seeking corporate sponsors before they enter production or even begin
talking to PBS.

Public television remains a fertile area for arts groups
already skilled at fundraising. Both locally and nationally, a
large share of public television production depends on the
entrepreneurial initiative of groups able to attract underwriters or sponsors for their programs. Quality considerations aside, delivering a program with sponsorship may be a
major incentive for public television to accept the program
and a major opportunity for groups to have their programs
aired.

In cases where exposure is the primary motive, groups may
be able to have their programs aired on public television if
they are already paid for and can be made available at no
licensing charge. In fact, they may have to pay an administrative charge to the public television station to cover the costs of
airing the show—costs which can run into thousands of
dollars for even a locally broadcast half-hour program.

Pre-Selling and Co-Production Deals

Perhaps the most complex but desirable way to finance a
production is selling the rights of a production before it is produced. Depending upon your perspective, this is called *pre-
licensing,* a *pre-buy,* or a *pre-sale.* Under such an arrangement, a
group receives the money, or a commitment of money, before
it has completed or even begun the production. This is one of
the most difficult ways to secure financing and one which is
most dependent on the attractiveness of the total package. In

some cases, though, the ability to attract a hot property or big star may be enough to get many buyers to write a check.

Outside the United States, the television market is growing so quickly that many new opportunities have been created for arts organizations and independent producers. Presently, the foreign television market is composed predominantly of countries which have only two or three television stations. In almost every instance, these are public television stations, either government-owned or closely regulated. However, commercial television is now burgeoning around the world. A new fourth channel in Great Britain is taking an active role in developing original programming and new channels, both broadcast and cable, are coming on stream in many other countries, both large and small.

Reaching these markets is a rather intricate and costly process. For those seeking to pre-sell rights, distributors or producers experienced in program marketing are much in demand. Many independent distribution companies regularly attend the international television markets and have helped form co-production partnerships between independent producers and television companies around the world. A *co-production deal*, as distinct from a pre-buy, usually involves the foreign (or domestic) television partner of the independent producer in the actual production, possibly contributing facilities, locations, or talent, as well as cash. For its participation, the partner may secure a license to telecast the program in its territory, the right to distribute in other territories, and/or a share of profits. Pre-buying, however, merely secures the rights to broadcast a program; in exchange for paying in advance of the production, a lower license fee is often negotiated, as compared with licensing the rights after the production had been completed. This is one of the major rationales for any foreign broadcaster to pre-buy rights. The decision to pre-buy rights is often based upon the attraction of one or more elements of the package and the likelihood of a superior production.

No efforts to pre-sell or secure production partners should begin without detailed budgets in hand. Production budgets break down into two categories: *above-the-line* and *below-the-line*. Above-the-line usually refers to the total cash costs of the talent and the key producing personnel of the production. In some cases, the producing personnel included in the above-the-line budget are only the producer and the director, or the executive producer, producer, director and certain associate producers. Above-the-line costs also include payments for rights, residuals, clearances, and all other expenses associated with the intangible properties of the production. The cost of crew personnel actually involved in the production is included in the below-the-line budget. The below-the-line budget summarizes the physical costs of executing the production — cameras, facilities, sets, post-production, lights, and all the other hard costs of equipment and physical services.

These distinctions between above- and below-the-line elements are worth keeping in mind because certain costs of a program can be targeted at certain kinds of investors or co-production partners. A thoughtful, accurate budget can serve as a road map in prospecting for financing. For example, if a certain amount of cash has been raised, it may be easier to secure a co-production partner who is prepared to make a facility available than to secure additional cash from another investor, especially if a major star or bankable producer is already committed. Many facilities face a certain amount of what is called *downtime* (periods when not in use). A foreign broadcast channel or a local public television production company is frequently pleased to have an equity participation or a financial interest in the production in return for the use of otherwise dormant facilities. Such a production can give a company or channel both promotional exposure and the chance to earn revenues from the success of the completed property.

The enormous perceived risks in creating any kind of cultural programming for television make it increasingly likely that any arts-oriented production will require many pre-buyers, partners, or co-producers to share these risks. Companies or individuals which contribute cash to a co-production deal usually put themselves in a position to recoup their investment first from gross receipts earned by the production. Recoupment means that revenues, whether from licensing fees or from additional investments, will flow back to the party which made the investment. Once a deal is structured, there will usually be an order of recoupment which sets forth the sequence in which investors are compensated from the production's revenues.

One alternative to elaborate orders of recoupment — it may also be a component — is what is called *pari passu:* each investor recoups his pro-rated share of the total investment from every dollar earned by the production. *Pari passu* recoupment allows every party to earn its respective percentage from the first dollar received, but it takes longer for each party to be paid back in full. Major investors with the greatest stake and leverage thus will often insist on having priority recoupment. Offering *pari passu*, while reasonable when there are many small investors of roughly equal shares, may not be practical in attempting to secure a major investor.

Rights and the Distribution Process

The intentions and terms of the deal extend into all areas of packaging, financing, budgeting, and distribution. In structuring some deals, these areas may be resolved simultaneously, while for other deals, some aspects may be contingent, and the deal as a whole will develop sequentially. Basically all deals have similar elements but different configurations, making it useless to attempt to set forth a sequence for ordering a deal which will apply in every case. What all deals do have in common is the purpose of serving as a blueprint of the rights in the production.

The key consideration in structuring a deal is control of rights. Co-producers or investors may request control of certain rights as well as the resulting income from sale of the rights, regardless of who controls them. For example, the sole repayment to a foreign broadcast station for investing cash or providing facilities, either as a pre-buy or a co-production, may simply be the license to televise the production in its territory for a certain period of time. The question then becomes one of establishing the fair market value of the television rights in a given territory, balanced in the deal against the value of a certain contribution to the production. Licensing revenues from major territories, such as the United Kingdom or West Germany, can be substantial. Since licensing is done on an exclusive basis, territory by territory, in a deal where a license is "thrown-in," a realistic sum representing lost licensing revenues to the production must be figured into the recoupment equation.

When rights in the production are reserved or assigned for marketing, their value becomes a function of the distribution process. Like the rights to televise the program, the rights to distribute it in various areas becomes an essential component of the deal. At the very outset of conceiving a production and imagining a deal, an arts organization must begin to envision the process of distribution. The executive producer or other individual responsible for putting the package together should contact distributors or have the expertise to answer distribution questions. Frequently, success in financing a program will depend upon the interest of a distributor in marketing it and the credibility of the distribution strategy and projections. The distributor's mandate is to maximize revenues from the program through the most profitable and efficient disposition of the rights.

The term often used to describe the segmentation of the sale of rights in the distribution process is *window*. A window is a period of exclusivity during which a property is sold to one and only one delivery system without being made available to another. A *delivery system* refers to the format of availability to the viewer — cable television, broadcast television, home video (video cassette and video disc formats), or even theatrical exhibition. A property being shown in movie theatres is in its theatrical window. During this window, the rights holder of the property may often be prevented from selling the property to any of the television delivery systems until the buyer of the theatrical rights has fully exploited the value of his property in its market.

A question that often comes up among arts organizations involved in television is how to retain maximum control over the property in both production (artistic integrity) and marketing (quality of exposure). Questions may arise, for example, whether to distribute to commercial outlets or to public television, which will not insert commercials. Similarly, should distribution focus on obtaining the highest revenues through limited pay cable licenses, or immediate, broad exposure on free television.

If it accepts a co-production deal, an arts organization frequently will be paid strictly for the production and forfeit all rights to market the program. In this type of deal, perhaps best-termed a *production deal*, a major program buyer or program service, such as HBO, the BBC, or even PBS, would pay an arts organization a sum of money to produce the program. Subject to negotiations, they will usually pay the group one-third in the pre-production stage after signing the deal, another third upon completion of principal photography, and the final third upon delivery. Although there are many intricate variations, this is the general principle of such a deal. In exchange, the program buyer may request all rights to the property in perpetuity and, after the total production costs have been recouped, will share the net profits with the group in some percentage which may be 50/50. Upon completion of the program, the arts organization will have no further control over how the program is distributed.

In fact, the program may not be distributed at all. The program buyer may determine that its quality or marketability is not sufficient to justify the expense of distribution. Or, he may decide that several other similar properties which he controls may have higher earning potential, prompting a holdback of the program from the market. The program buyer's responsibility to the arts organization producing the program has been technically fulfilled by providing the needed financing.

Arts organizations may, consequently, find the marketing of their program unsatisfactory. One option is to design the deal so that the rights revert to the arts organization under such circumstances. Any organization that gets to this point in a deal will already be working with lawyers, distributors, and various professionals who can analyze the ramifications of a deal to ensure that the organization's interests are protected.

Tasks and Strategies of the Distributor

Without distribution there is no pay-off on the deal and no substantial benefit to the producers. Often, in putting the deal together, the entities asked to contribute cash or facilities (as licensees, co-producers or investors), will ask for some indication of the marketability of the product — letters from distributors (if not directly from end buyers or prospective end buyers) may serve to establish that the property is commercially viable and has market potential.

At this point, a distributor may become involved in the packaging and even the financing of the production. Many of the most successful producers today began as distributors. After learning what kinds of properties had potential for sales in the market, they began to invest in projects before they were produced, in effect becoming producers, even though many had never laid hands on a camera nor could distinguish the look of film from tape. Such marketers are possibly the best-rehearsed executive producers, the most pivotal players in putting together a package. But whether or not the distributor is involved in the financing or the producing, he should be an expert in all of the possible windows of distribution and should be conversant with all the present and future technologies of delivery, since new opportunities keep opening up: video cassette and video disc, pay television and cable, as well as pay-per-view, low-power television, direct broadcast satellite, not to mention the traditional means of "standard television" delivery such as network broadcast, broadcast syndication, and so on.

A distributor with a track record will have had considerable experience in selling properties in all markets worldwide. If a distributor who has this breadth and depth of resources cannot be attracted, the arts organization may need to attract a group of distributors, each of whom may be smaller but may specialize in one aspect of the marketplace or one type of delivery system. For example, there is a new category of distributor which specializes in new technologies (having formed relationships with the home video, pay television, and cable companies). There are also distributors who specialize exclusively in the foreign marketplace; their metier is to take domestically-produced programming and offer it overseas.

Distribution has become a much more complicated and

vibrant area in light of the expansion of television delivery and the proliferation of new technologies all over the world. In the past, distribution has been dominated by the major syndicating companies which specialized in selling reruns of off-network programs to independent broadcast stations. Companies like Viacom and Worldvision, as well as the television sales arms of the Hollywood studios, have been the leading players. But now, myriad companies have entered the industry. Often they have the entrepreneurial energy and intensity of commitment to smaller projects that arts organizations find attractive.

At some point, most producers or executive producers will bring a distributor into the discussions with the arts organization about the design of the deal. As the architect for the program's marketing, the distributor should work with the producers of the arts organization's program to develop a strategy which will encompass all possible outlets for the program and seek to maximize its earning potential. This strategy should map out a *pattern of distribution,* meaning a sequence of sales for a program through every potential delivery system (window) and territory. For example, a program produced in association with public television in the United States may be designed for a broad national audience to watch on PBS. But a distributor who is involved from the outset might recommend that the deal be negotiated to provide a premiere of the program on a cultural cable service.

The reason for this is simple. Cable services may demand exclusivity—the right to premiere their programs—since people are paying for the right to watch these services, whether for their basic cable hook-up or an added fee for a premium service like the Disney Channel. If the program airs first on free television over PBS, a basic cable or over-the-air pay television service may have no incentive to license the property. Therefore, one lucrative outlet for distribution might be lost. Negotiations, however, can be pursued with PBS in such a way that the program has a cable window first, which at best can reach only a fraction of the total population since cable is in only 35 percent of American homes. After a certain period of exclusivity, the program can have national airing on PBS. At that point, no continuing runs may be made available on cable. After the PBS window, the program may revert to the cable service, or the program may run concurrently on PBS and cable. Securing the premiere window is crucial to the cable service, and they may be prepared to pay as much to reach 10 percent of the American public as public television is prepared to pay to reach virtually 100 percent.

Similarly, if the cultural property had potential in the home video market (which is doubtful since the video cassette and video disc market have typically been driven by feature films and popular performance programs, rock concerts, etc.), the home video company would demand a window which preceded delivery on either free or pay television since many people who own video recorders would tape the program rather than buy or rent the cassette. The sequence in the optimal pattern of distribution would provide release of the program in video cassette and video disc first — for 90 days to 120 days — after which the program would be made available to pay television. After pay television, which traditionally demands a window between one and two years, the program would be made available to basic cable, national broadcast over PBS, or through syndication. It is possible for programs to rearrange this whole pattern and move in circuits where there are different progressions of markets and audiences. But any transgression from the logic of reaching increasingly broader audiences requires special negotiations. Any arts organization which gets to this point wil certainly be working with distributors and consulting other experts.

Patterns of concurrent distribution present no exclusivity problem, however, when the rights are marketed by territory. Licenses may be sold simultaneously to numerous foreign territories without conflicting or impinging on any of the rights sold domestically. In most cases, the windows for those rights are sequenced within each territory in the same pattern as in the United States. For example, in England, the home video rights would be sold first, followed by pay television rights (if pay television were a significant factor in England, which it is not), and then broadcast rights. As mentioned earlier, the main consideration is the division of home video and broadcast to avoid the possibility that owners of home video recorders would tape the program off the air, undermining the potential sale of the program on prerecorded cassette or disc. (Home video companies argue that they must have protection from other forms of distribution until they have had a chance to exploit the program in a way that asks the viewers to pay the most for the program in order to get it first.)

Pay-Outs and Patience

An important feature of the final deal is the framework for payments, or the *pay-out*. Often, when all the elements of financing are in place, there still may not be enough money to meet the demands of all parties. Consequently, certain parties may be asked to defer their fees. The star, for example, may be asked to take less than a promised $100,000 salary in exchange for "points" or what is called a *backend position*. Instead of being paid up front, the star, the producer, or the arts organization itself — any party in the production — agrees to a postponed payment out of revenues expected to be earned by the production. Backended parties usually join the investors on the recoupment queue. If the sales of a program fail to meet distribution projections, any backend position, whatever the percentage, may amount to no position at all. Furthermore, the parties contributing hard cash to the deal are usually in first positions to recoup that cash, so the production facility (which may have deferred costs for downtime utilization), the star (who may have given a performance more as a contribution to the arts organization), and the arts organization itself (which will benefit promotionally from the television exposure), may all be argued into backend positions near the end of the revenue stream.

Backend money can only be earned when sales of the program, after distribution fees, net revenues beyond recoupments of the principal cash contributors who are usually first in line. In cases where the major investor in the program is a program service, such as a pay television network or foreign broadcaster, the value of its license to televise the program may be allocated as an automatic recoupment of the entire production cost. This important point has been raised before. When Showtime pays $200,000 to finance a production, it receives the right to present that program to its audience without paying any more. In fact, it might have had to pay more than $200,000 if it were to have licensed that program in an already finished form. Therefore, all sales revenues from point one, even if they are sales made by that program service functioning as the distributor, immediately go to the backend of the production, since the costs of the production have already been recouped in the value of the program service's own license. Achieving this "end-user" financing is most desirable because it is also an automatic first sale. The quicker the production money is recouped, the sooner those with deferments, or with points as a "kicker" beyond fees, will see their backend in the bank.

Other pay-out factors delaying recoupment are payments which must be made to unions and rights holders for certain residual and clearance fees triggered by the different windows of distribution. Virtually all of the major unions and guilds have contracts with fee structures that are applied as the program is sold into different markets, the principle being that the unions and all the personnel in the production should be compensated for their roles as the exploitation of the property proceeds.

In the area of music, the fees that music publishers demand for home video rights are still being formulated. One of the major problems involving musicians' unions and music publishers is that no one is really quite sure how the new video technology will affect the value of a musical property. In some areas, where final guidelines have not been settled, each production may require a new negotiation with the formulas reinvented to accommodate the perceived value of a particular property.

Of course, there are solutions within these situations. The best advice for any arts organization is to find a lawyer, a producer, or a distributor who knows all the ins and outs of negotiating with rights holders and unions to insure that the program will be affordable to sell in certain markets. There are some cases where residual payments to the guilds and rights holders exceed the revenues which can be generated from selling the program into the markets which would trigger those payments, making certain sales impractical and certain levels of recoupment unlikely. All of these matters must be taken into consideration before projecting a backend payout to investors or co-producers.

General Considerations and Special Concerns

Much ground has been covered in charting the progression from ideal to deal and the realization of the deal as the so-called property. In addition to packaging the property, negotiating the deal, financing the production, producing the program, and securing distribution, not necessarily in that order, there are some general considerations of which an arts organization should be aware. These may have much to do with an arts organization's future success and its future work in television.

An arts organization should meticulously research all possible costs of the production so that reliable projections can be made. Some costs are not immediately apparent, such as the various fees which have to be paid to the middlemen involved in putting the deal together. From actual production to the point where the program is bought, there may be a chain of numerous middlemen or middle entities which serve as various sub-agents, sub-distributors, brokers, and licensees. A program which earns a million dollars in gross distribution revenues might return less than one-third of that to the basic producer or the arts organization after all the percentages and fees have been deducted by packaagers, distributors, agents, and all the parties involved in marketing the property. Marketing is a very expensive and difficult process. It is perhaps most difficult in the area of the arts because the programming is specialized and traditionally does not have broad commercial appeal.

Another important consideration is the term and conditions of the deal. All of the arrangements with the buyers, co-producers, and distributors have a fuse on them. The fuse may be a function of the amount of money earned by the property and/or an absolute amount of time. Sometimes, a certain amount of money must be earned or paid by a certain date for the term of a license, or the term of distribution control over a property, to be extended. If a distributor does not achieve a projected level of sales, or make up for it in the payment of a guarantee by a certain date, all the rights may revert to the arts organization or to the producer. Similarly, questions of exclusivity and nonexclusivity in the distribution of a property, or the right to transmit it, have to be agreed upon in negotiations before a deal is signed.

A special area of concern for arts organizations which produce programming is the use, control, and ownership of a program once it has been produced. This applies to groups which are producing on the local and regional level as well as to groups involved in national or international commercial ventures. The question of control relates directly to distribution, because the way in which a program winds its way through the marketplace determines the level of its exposure and the image projected of the producing arts organization.

At this point, the relationship between the arts organization and the distributor of the program is of critical concern. If a co-producer of the program gains distribution control, they need not distribute it themselves, but may select the distributor and earn a fee on top. In any case, an arts organization should have some say in the selection of the distributor and, possibly, some input into how the program will be marketed. Presumably, an arts organization has professional interests beyond producing programming for television, but the way their image is processed and positioned by television may have a great deal to do with their survival and success in live performance.

Reasonable control and approval of the marketing of the program should concern an arts organization as much as the execution of the actual production. In some cases, the arts group may exact a degree of direct control over distribution which can put the formal distributor in a quasi-nonexclusive situation. This can be problematic since distributors who do not have exclusivity have less incentive to market a property well. In fact, they may find themselves in competition with the arts organization that wants to market the property themselves.

This problem can be overcome by negotiating an arrangement whereby certain areas are withheld from the distribution contract and specific guildelines are set forth in other areas. For example, a dance company that travels thoughout Europe and has already been taped by foreign broadcasters (perhaps they have done a television show for West Germany or developed a relationship with the BBC, may insist in negotiations with the distributor that sales to those particular buyers are not commissioned at all, or at full fee to the distributor. On the other hand, it may be in the arts organization's best interest to turn those contacts over to the distributor; indeed, established relationships with potential buyers can be used as an enticement to attract a very strong distributor and a particularly attractive distribution deal. In addition, the potential of pre-sales, or any high probability of sales, even before a distributor is brought into the picture, may put the arts organization in a position to secure some type of distribution advance, or guarantee.

Finally, an arts organization should keep in mind that the whole endeavor of programming for television, from business wheeling and dealing to production nuts and bolts, is an intense but elusive learning opportunity. Once the wheels start turning, the lawyers, producers, distributors, and other agents may wrest most of the execution from the arts group's control. But no production can go forward successfully without superb communication among all parties and partners in the process. If the foregoing discussion serves as nothing other than cue cards for an arts group taking first steps from stage or frame to screen, the chances of success may not be enhanced but, hopefully, the process has been made moderately more accessible.

The Concepts of Negotiation

by Timothy J. DeBaets

This chapter is intended to give a newcomer to the cable field a basic understanding of the concepts involved in negotiating a deal for a cable television production. It assumes that the deal-making described in the prior chapter has taken place. First, this chapter discusses general negotiating conditions. Then, it discusses the meaning of terms likely to be found in the final written agreement and alternatives to those terms for which you should negotiate. The actual negotiation process will take place between the lawyers and will consist of telephone or in-person sessions. Certainly, what follows cannot be taken as the last word on the subject; the world of cable television continues to be marked by constant and dramatic change. Timely, expert legal assistance will always be needed. But with this information, you should be able to evaluate more effectively the assistance you receive.

You should also begin to read the trade publications, which will acquaint you with the cable marketplace and its prime movers and shakers and keep you up-to-date on a number of issues relevant to cable negotiations, such as licensing fees, programming deals, and the shifting fortunes of the various cable companies (the demise of CBS Cable and The Entertainment Channel, for example). *Cablevision, Channels,* and *Electronic Media*, for example, contain substantial information about the cable field. *Variety* is must reading for all entertainment fields, including cable television.

The next step is to retain a lawyer before negotiations begin. The cable television company and/or independent producer will have a lawyer. As a newcomer to the cable field, you are likely to be taken advantage of without a lawyer to protect you.

Whose Side Are You On?

Once negotiations for a cable program have begun, it is necessary to be sensitive to conflicts of interest. The following hypothetical situation illustrates how complex things can get.

There are four parties involved, each with its own interests to protect: a summer theatre festival, a mime troupe, an independent producer, and a cable programming service. [1]

To augment its forthcoming season of new plays, the summer theater is considering a number of special presentations, including the mime troupe. Through extensive touring, the troupe has received widespread acclaim — for one program in particular — and the festival managers feel that their presentation of the mime troupe's best-known progam might interest the cable programming service. Already, the festival has been approached by an independent cable producer, who is interested in taping one, two, or maybe even three events — under the name and auspices of the festival — which he would then license to a cable service. Together, the independent producer and the festival management must decide which of the summer's special events would be the most attractive to cable, so that the producer — perhaps with the assistance of the festival — can begin his talks with the programming service.

At this point, the interests of the theatre festival and the independent cable producer converge: The producer wants to make a profit from the licensing; the summer festival wants to profit from the live performances *and* the cable presentation. The summer festival might even seek a long-term arrangement for the taping of future presentations, which could provide an ongoing supplement to box office revenues and contributions.

The mime troupe, on the other hand, is in a different position. It may never perform at the festival again, so the performers will have little interest in the festival's prospects for future cable deals. Certainly, however, the troupe's future will be affected by the taping: It is very likely that under the terms of the television deal, the company will not be allowed to perform its most popular work in other media for some period of time. Consequently, the performers will want the most favorable terms possible in the one cable deal in which they are participating.

Should the summer festival negotiate on behalf of the performing group? Should the festival negotiate strictly with the independent producer and allow him to make the deal with the cable company? Should the performing group negotiate for itself? With whom? The producer? The cable company? And what about the other presentations the festival might want to include in its package of cable offerings? What if they include one of their own plays? Should the festival negotiate on behalf of the author and actors, or should the negotiations

---

1. Throughout this chapter, the term "artist" will be used. It is meant broadly to apply to an individual artist or to an organization or group of artists.

be left to the artists' agents and lawyers? Who does the author's agent talk to? The festival, the producer, or the cable company?

Obviously, the negotiations could proceed along a number of different lines. Whatever course is chosen by each party, everyone must be sensitive to potential conflicts of interest and to the fact that, at this point, the various parties have both converging and diverging interests. For instance, a long-term arrangement between the cable company and the summer festival for the licensing of future productions might result in a smaller payment to the mime troupe for their one taped performance. The performers may be able to demand more money if everyone involved is interested solely in their show. The independent producer may have a number of projects in the works with the cable company, and for the sake of his ongoing business relationship, he may be willing to make certain concessions in the deal for the mime program. He may even have a commitment to the cable company to produce the program as cheaply as possible, something which would not be in the best interests of either the festival (if this is a one-time project) or the performers.

None of this is meant to imply that everyone involved in a cable deal cannot be treated fairly, only that each party must be responsible for seeing that his own best interests are protected and promoted during the negotiations. It is advisable that the summer festival and the performers be individually represented by an experienced agent or lawyer. It may be assumed that the independent producer will either have a lawyer or be able to negotiate properly without one. And certainly, the cable company will have an attorney — or, more likely, several.

At this point, the use of in-house lawyers by cable companies should be briefly discussed. Most often, the first attorney you meet will be a member of the company's business affairs department. His responsibility is to meet with the other parties and negotiate the terms of the deal. He may also have been involved in *initiating* the deal, either through an independent producer or directly with the performing group or theatre management. Once the terms have been agreed upon verbally, the business affairs lawyer will turn the matter over to the cable company's legal department to draft the written contract. The legal department lawyer, in drafting the actual document, will continue to negotiate, but his emphasis will be on "legal" points (such as whether disputes should be submitted to arbitration) as opposed to "deal" points (such as the amount of the licensing fee). Not all cable companies are organized this formally. Some will have the same lawyer — or team of lawyers — negotiate the deal and draft the contract. Still, if a cable company has at least one, and more likely two, sets of lawyers, this should alert you to the need for at least one lawyer on your team.

When all parties are individually represented and promoting their own interests, it may be more difficult to arrive at a final agreement, since lawyers and agents are generally contentious — that is their job. It is also their job to see that their clients are not taken advantage of. With individual representation for all parties, it is more likely that the final deal will be fair to everyone involved. In the brief history of arts-cable programming deals, experience has shown that if there is an imbalance in the final deal, someone is likely to be unhappy, and sooner or later, the repercussions will make everyone miserable.

Early on, the relationships among the parties must be clarified regarding who is responsible to whom and precisely what each party is obligated to deliver. Most often, the funds to produce the program will be provided by the cable company. Consequently, it will want to have a contract with everyone involved in the project. In the case of the mime show, the cable company financing would mean that the programming service would draw up a separate contract for the summer festival, for each performer in the mime troupe, and for the independent producer. Each of these contracts must be reviewed and negotiated separately by the party concerned.

On the other hand, the summer festival could produce the program itself and eliminate the independent producer from the deal. In that case, the festival would first negotiate with the performers, then make a licensing deal with the cable company. The disadvantage to this is that the festival would substantially add to its responsibilities — both to the cable company, to whom the festival would be obligated to deliver a finished product, and to the performers, whose financial compensation for the taping would become the direct responsibility of the festival. If there are residuals or net profits to be paid, this would put a long-term burden on the festival's accounting staff. And if the festival is to function as a cable producer, not only will extra staff be needed, but the energies of the entire staff — throughout the season — may be consumed in the exigencies of television production.

Suppose that the independent producer wants to produce the program "on spec" — that is, without a pre-sale to a programming service. The independent producer would then put up the money (perhaps supplemented by an allocation from the festival's budget) to tape the program. The problem with this kind of arrangement, from the point of view of the festival and the performers, is that they might not know the terms of any subsequent deal made between the producer and the cable company. Although it is possible that the mime troupe could become involved in future negotiations, typically, the independent producer will want to own outright all rights necessary to sell or license the program. He will want to know the entire costs of the production in order to determine an appropriate licensing fee. Consequently, he will seek to conclude all negotiations before the taping, in particular so that no one will be able to hold out for better terms after he has spent considerable money on the production and/or is close to a deal with a cable company. This allows the producer to make the best deal possible—from his perspective—without any input or interference from the other parties. From the performing artist's perspective, then, productions on spec are less desirable; it is preferable that your show be financed and produced by a reputable cable company with whom you can negotiate directly and make the best deal possible for yourself.

The hypothetical deal that most of the above discussion is based on involves four parties: the summer festival, the mime troupe, the independent producer, and the cable company. Some projects are even more complex. For example, if one of the festival's plays were part of the package, a new player would enter the game: the playwright. And if the show were a musical, there might even be three separate authors—a composer, a lyricist, and a librettist—and a separate agreement would have to be reached with each of them for the cable rights to his contribution. A project can also be much less complex than any of these examples. For example, the cable company could arrange directly with the mime troupe to tape a performance in the company's own television studios. Neither the summer festival nor the independent producer would be involved. Obviously, a one-man show performed by the creator of the material would be even simpler. For purposes of clarity and simplicity, the following discussion of written agreements will focus on these less complex situations.

The Deal Memo

Often, a written agreement with a cable company is arrived at in two stages. First comes the *deal memo*, which can be as simple as a letter from the cable company to the artist. Even though it may be as short as two pages, it will cover all the essential points regarding compensation (the artist's fee and his percentage of the net profits, if any) and credit (how his screen credit is to be worded, where it will appear on the tape, and for how long, etc.). Clearly, the dates for taping the show should be indicated. If the cable company is to have an option with the artist for future programs, this should be spelled out in the deal memo.

Protection of the artist's rights to his work, including his copyright, is another critical component of the deal memo. There must be a clear statement of which rights the cable company is licensing and which rights the artist is retaining. If the cable company is acquiring only the rights to present the work on cable, for example, then the artist's reservation of rights (for other media and for live performances) should be specifically and explicitly expressed. Sometimes, the cable company will want a representation and warranty that the work is original, and indemnity if this proves not to be the case; sometimes this will be left to the final contract. In any case, such provisions should be made mutual.

The deal memo is usually drafted and signed by the cable company (sometimes it is countersigned by the artist) and is a binding agreement. Frequently there is no other contract beyond this. In fact, it is sometimes to the advantage of the artist or performing company *not* to sign an additional contract, because the more detailed document will often contain representations, warranties, indemnities (mentioned above), and other protections that are mostly for the benefit of the cable company.

A final contract will be necessary, however, if the artist is to receive a percentage of the profits. In that case, the numerical percentage should be stated in the deal memo, although the precise definition of just what the artist is getting a percentage of will be left to the final contract. Almost certainly, the artist's percentage will be based on *net profits*, the earnings left over after the cable company has recouped its investment. This may sound like a perfectly straightforward concept, but the artist should be aware that the company will count more than production costs in its investment figure, and the figure will continue to grow long after the actual taping is completed. (Advertising is one of the most obvious post-production expenses incurred by the cable company. There are a number of other, less obvious ones, all of which will be discussed later.) Unfortunately, there have been so few net profit deals with cable companies that it is difficult to know whether profit participation can be a meaningful form of compensation for the artist.

Often, once the deal seems likely, both the cable company and the artist will commence pre-production even before a deal memo is signed. The cable company may even be willing to advance part of the artist's fee or certain production expenses. Obviously, the company will be anxious to have the program completed, and the earlier things get moving, the better. But for the artist, this can entail certain risks. Once he has become professionally and emotionally wedded to a project, it can be very difficult for an artist to walk away, even if the cable company fails to give him the support — and the deal — he had initially expected. And most likely, the cable company will be very aware of the artist's deep commitment. If a dispute arises before there is any written agreement, the cable company's business affairs person will keep saying to the artist's lawyer, "Well, you know this deal is going to happen. This term is not a deal-breaker[2] — we just can't, as a matter of policy, give you such and such."

Good contract terms are difficult enough to achieve for an artist, especially for lesser-known artists and companies. Typically, artists receive low initial offers in the cable television field. Cable companies do not want to spend large sums of money on programming: They need volume, but they may not necessarily care about quality. Nonetheless, the lawyer representing the artist is under an obligation to get the best possible terms for his client, and if the project is already into pre-production before the final terms are ironed out, negotiation becomes much more difficult.

In such a situation, tough as it may be, it is sometimes advisable for the artist simply to stop working with the cable company and go on to other projects. Naturally, this will heighten the hysteria of the cable people, and negotiations may resume amid a lot of tension. But if the artist refuses to take such a stand when necessary, he will have delivered himself right into the hands of the cable company, and his attorney will be powerless to protect him.

Think of a deal memo as having only the basic core elements, the skeleton of the actual deal — those items that come immediately to your mind without the need for your lawyer to explain what separate card credit means. You will still need a lawyer for the fine points, of course, but a deal memo contains the *sina qua non*. It is also the first test of your relationship with the cable company, a relationship that you may have to live with for a long time. If you cannot agree on the basics of the deal memo, you should not do the program.

The Contract

Independent producers often attempt to dispense with the deal memo and present artists with a final contract which may

2. By *deal-breaker*, we mean a negotiating point which is so important that if it is not conceded by one side or the other, the parties will not complete the deal.

be only several pages long. However, if the contract you are about to sign has been drafted by a cable company, it is likely to be a much lengthier document, possibly as long as twenty-five pages. Either form is acceptable if the artist is properly protected.

Generally, the contract will be divided into two parts. The first part will be an expansion of the deal memo. It will include a more precise definition of terms, but otherwise will be fairly straightforward. The second part is a different matter. Often denominated as "Standard Terms and Conditions" or "Additional Provisions," this section of the contract is what is known as the *boiler plate,* and typically will be unreadable. It will also contain some of the most critical provisions of the agreement, so you cannot ignore it. Sit down with your lawyer and carefully read through the entire document, making sure that you thoroughly understand every single point.

The contract used by the cable company will most likely consist of numbered paragraphs. The first part will begin by reciting the names of the parties and the services to be performed by the artist or performing arts company. It may state that the artist is an employee of the cable company if the artist is a performer. If the artist is merely licensing artistic work, there is no need for him or her to be an employee, since this may suggest that the cable company owns the artistic work. The beginning paragraphs will also describe the services to be performed by the artist as well as the dates during which the services are to be rendered.

Next will follow a discussion of rights. In the case of an author, the cable company will usually purchase a license of the rights — a specific right to use the work for the limited time and purpose stated in the contract — but will not buy the work outright. In the case of a not-for-profit organization which has produced the work, the cable company will license the production, for which the producing organization will be paid and given credit. Whatever rights are granted to the cable company, it is important that they are limited to use in connection with the program. Care must also be taken to insure that underlying rights (such as an author's copyright) are preserved for their owners. Of course, these points should have been part of the deal memo, but they must be restated in the final contract.

The next several paragraphs of the first part of the contract will discuss compensation. If there is a license fee, that should be stated. If there is to be some form of profit participation, the percentage should be indicated at this point. (The definition of net profits — if that is the basis for the participation — will usually be reserved for the standard terms and conditions.) Expenses such as round-trip transportation, accommodations, and per diems should be included. If the artists have to travel to tape the program, and if the cable company has agreed to pay for any other personnel (staff members, spouses, etc.) to attend the taping, this should be stated here. Apart from travel expenses, there is likely to be a clause stating that the artists will not be reimbursed for any other expenses unless they are approved as part of the production budget or by a program production manager.

Sometimes, if there is a stalemate concerning the terms and benefits to be included in the contract, it is possible to negotiate that some items be included in the production budget. Items which are of critical importance to the artistic quality of the project may be more important to the artist than the actual fee he or she may receive. For example, a dance company may try to include at least the following items in the production budget: an assistant director to work with the choreographer; additional television monitors for rehearsal sessions to help translate the dance from a live, stage production to tape; payment for rehearsal space, which may also be used by the choreographer for work unrelated to the cable program; payment of dancers' salaries in advance to relieve the company's cash flow problems; and payments for rehearsal clothes and shoes, which the dance company will be allowed to retain after taping. These and similar provisions can be part of a cable deal without being spelled out in either the deal memo or the final contract, as the cable company's policy will sometimes prevent assigning these over in the contract. Since you would be essentially receiving an informal commitment, prudence would indicate that extremely expensive items should be expressly provided for in the contract or deal memo.

In addition to a license fee, compensation for the artist may include a fee per play-date. This could literally mean each time the program is aired, but normally a play-date consists of a number of presentations within a specified period of time, For example, an entire week, during which the program may be presented an unlimited number of times, might be contractually defined as a single play-date. It is also possible to negotiate for a limitation on the number of times the program can be presented during the specified time period, after which the artists must receive additional payments. For obvious reasons, cable companies tend to resist the concept of fees-per-play-dates. Apart from the additional cost, it makes the accounting process more complex. Generally, this is an item which only well-known artists can negotiate. If you succeed in getting it, more power to you.

Even if you are to be compensated for play-dates, remember that a cable company is not obligated to air the program. This is known as the *pay or play* requirement, meaning that the basic compensation will be paid to the artist whether the program is aired or not. Although you may want a guarantee that the program on which you spent so much effort will be cablecast, it is only reasonable that the cable company be allowed to decide what it will air, since the company alone is ultimately responsible for its programming choices — to the public and to the regulatory authorities.

There may be another discussion of rights at this point in the contract, wherein the cable company indicates that it owns the copyright of the videotape and has the exclusive right to use the videotape in whatever territory for which it has been licensed. There are two significant provisions here: copyright ownership and exclusivity. With respect to copyright ownership, remember that it is possible for the videotape to be copyrighted separately from the underlying work. It is vitally important that the artist reserve the copyright of the underlying work in order to continue to own it and use it in other media. Under the separate copyright arrangement, the artist has contractually licensed use of the work in a certain medium; he has not sold it outright. The artist may also argue for ownership of the copyright in the videotape. However, if the cable company has fully financed the program, it is likely to insist upon owning the copyright in the videotape.

In most cases, a cable company will insist on the exclusive right to use the videotape on cable television, which means that the artist, therefore, cannot enter into a contract with another cable company for the rights to this particular program. This is a reasonable request. It would be of little value to a cable company to license a non-exclusive production of a Broadway show, for example; a competitor could present a different production of the same work and draw viewers or subscribers away. It is possible, however, to prevent the company from licensing the tape to broadcast television — network, local, or syndicated. Certainly, if the cable company *is* to have the right to authorize use of the tape in other markets, additional compensation for the artist — payable in the event that such sales are made — should be negotiated into the contract.

You may wish to grant exclusivity in cable and allow the cable company to sell the particular program to other noncable media. The cable company may insist on this right, which it may be reasonable to grant. For example, distribution to European television may not be possible for an individual artist or even for a substantial not-for-profit entity. Allowing the cable company to distribute the program to noncable media, both here and abroad, may be useful if the artist does not have the means himself, and it can be done subject to his profit participation.

The contract will also include certification that you, as licensor, represent and warrant that the work is original and owned by you, or that you have the right to license the work.[3] This is perfectly legitimate. You cannot sell something you do not own. To maintain your own integrity, you must insure that you are licensing original material or that you have obtained the rights to the work. In conjuction with this requirement, the cable company will require that you indemnify and hold it harmless if the material is not original or if it turns out that you do not have the right to use it. This, too, is appropriate and usual, but this provision should be made mutual — you should require that you receive indemnification and are held harmless (including requiring the cable company to bear the cost of your attorney's fees) in the event that you are sued for that part of the program which is within the province of the cable company — for example, narration provided by the cable company or contributed by a different artist.

The contract may also have a clause allowing proration of payments in the event that the program is postponed due to *force majeure* (matters out of the control of the cable company, such as wars, riots, or intergalactic implosion). Agreeing to such a term may not be in your best interest, but it is often academic. By the time the final contract is signed, the artist may have received most of the payments due, or the program may even be completed. Not infrequently, a cable company's legal department will be so far behind in drafting final contracts that you may be well into the editing process before the document is completed. Even on a network series, a production company may be operating with nothing but deal memos with its creative staff (such as writers and artistic director), and if the series is cancelled after only a few shows, the company may not even bother with full-length contracts. In the event that the contract contains a force majeure provision and the actual production work has not begun, you may want to resist agreeing to it; if the production is well underway or completed, it becomes a minor point.

There will be a discussion of credits in the contract. This is a critical matter, and you must be very precise as to the type of credit you want to receive. We have all seen television programs and films with the author's name or a star's name over the title of a work. This does not happen often, however, and it is an obvious indication of both the artist's clout and his agent or lawyer's ability to negotiate credits. Try to argue for *full-frame* or *separate card* credit, meaning that your name or

---

3. Of course, this does not apply to the artist who is solely a performer.

your performance company's name will appear by itself on the screen, with no one else's name appearing simultaneously, and in size as large as all other credits. You may also indicate placement of credits—whose name will appear first, for example. In conjunction with the credit provisions, there will be an authorization to the cable company for the use of the performer's name, likeness and biographical material. This is a standard request, but it is suggested that the biographical material be provided by the artist.

Considerable cable programming is produced without being subject to union agreements, as a preceding chapter explains. However, there will be a clause in the contract requiring the artist to join the applicable guild or labor union if necessary. This provision is reasonable. Because there currently are ongoing negotiations between various cable companies and the entertainment guilds, a guild agreement could be signed during the production of your program requiring the cable company to hire only union members, or provide that non-union members join a union for that program. (It is not likely that a guild agreement signed after your production is complete, or after the program is broadcast, would require guild membership.) It should be noted that, in the absence of union agreements in much of the cable industry, many of the protections provided by these agreements for creative artists (and hence additional expenses for the producing entity) are never employed.

Near the end of the first part of the contract, there will be paragraphs dealing with issues which also are frequently covered in the boiler plate, or "Standard Terms and Conditions." There will probably be a paragraph providing that all notices—both for the cable company and the artists—be sent to the specific address listed for each party in the contract. As an artist, you should add that a carbon copy of any notice to you should be sent to your attorney as well. In case of a dispute, both you and your attorney will be simultaneously notified and will be able to respond in a more rapid fashion.

There will be a paragraph specifying that the artist will not accept money in violation of the Federal Communications Act. Certain provisions of this law were enacted in response to scandals in the television industry, such as the rigged quiz shows in which participants were supplied with answers to questions. The provisions are meant to prevent similar deceptions of the public. The paragraph is a standard term to which you should have no objection.

There will also be a provision requiring that a waiver by one party of any term or condition shall not be deemed a waiver of any other term or condition, nor of any subsequent breach. This is to the benefit of both sides, and its acceptance is recommended. Further, the contract will most likely provide that the invalidity of any one provision does not affect the validity of other provisions in the agreement. Again, this is logical and reasonable, and there is no reason to object.

The contract will indicate that it is governed by the laws of a particular state. This provision is necessitated by the legal variations from one state to another. In the absence of such designation, the court would have to determine which state law applied. Invariably, the cable company will insist upon the state in which it is located. You may argue for your own state if it is a different one, but the cable company is not likely to bend on this point.

One way to protect yourself in case of a dispute is to provide that all disputed matters be submitted to *binding* arbitration before the American Arbitration Association, which is based in New York but has offices nationwide and can arrange for a hearing with arbitrators in any city. Arbitration is a quick, simple, and cheap way to resolve disagreements. Although a cable company may be able to afford a costly lawsuit, artists and arts organizations will have fewer resources and will benefit from the arbitration process. Unfortunately, the cable company — aware that the prospect of an expensive litigation may cause an artist to settle a suit prematurely or to decide not to file a suit in the first place — may resist this provision. Still, it is worth arguing for.

The contract will also typically provide that the cable company can assign its rights — that is, give to another party all rights and interest in the contract. It should be noted that the right to assign is usually not mutual. The cable company wants the artist's particular skills or artistic work, and will not allow assignment to another artist. This is reasonable. If CBS Cable contracts for the services of Twyla Tharp, it certainly will not want her to assign her obligations to another choreographer. The cable company will also insist on the right to assign in the event that it ceases operation, such as CBS Cable or The Entertainment Channel, and wants to sell or license its programming to another entity. The cable company may also reorganize itself and assign its programming to a new corporation. The artists should be protected by a provision that holds the cable company responsible for the obligations to the artist notwithstanding any assignment.

Finally, there will be a provision stating that the agreement is the entire understanding of both parties and cannot be changed except in writing signed by both parties. This is quite sensible. It precludes the existence of other oral or written agreements which may be contradictory to this final agreement. However, before signing a contract having such a clause, take a careful look at your deal memo to make sure that everything in the deal memo is incorporated into the contract. You could also add a provision incorporating the deal memo by reference, although normally the final contract will supercede everything in the deal memo.

Standard Terms and Conditions

Now, on to the boiler plate. The standard terms will contain provisions that amplify those outlined in the first part of the contract. For instance, if the cable company is entitled to assign rights to the program, there may be a clause in the standard terms and provisions allowing for outright sale of the program. There may be a clause affirming that the cable company owns the program exclusively, not withstanding any profit participation which the artist may have. Thus, a profit participation is a contract right and does not give the artist any right of ownership in the film or videotape. These are acceptable provisions if, as discussed above, the boiler plate also includes that any sale or assignment of the program must be subject to all the rights in your contract with the cable company, and that the artist is notified of any such sale. It would be desirable for the artist to have some control over to whom the program is sold (like a baseball player having the right to control to which team his contract is sold), but this is a term you are not likely to receive unless you are in a remarkably strong bargaining position.

There is also likely to be an exercise-of-rights-to-the-program section, which will expand on the pay-or-play clause in the first part of the contract. Typically, it will provide that the cable company is not obligated to "exploit" the program, i.e., to telecast it on cable television or to sell it to other media, and that the company has not made any representation or warranties that it will do so. The contract will also state that the cable company has not represented that any particular amount of money will be received as net profits. This is standard language. So far as compensation goes, you should think in terms of the licensing fee that you receive. You should be happy with this fee, and realize that net profits or fee-per-play-date revenues may or may not materialize. You should be properly skeptical about such matters and look at the overall deal and the program itself. Are you happy with the immediate compensation? Does the program give you appropriate credit and demonstrate your artistic merit? You cannot count on post-broadcast compensation formulas; there is not enough experience in the field to know what sort of income can be realized from them. And there is no indication that cultural programs will have the kind of success enjoyed in television syndication by popular series such as "M*A*S*H" and "I Love Lucy." If you are not happy with what you are receiving up front, you simply should not make the deal.

Still, quite frequently, participants in a cable program are offered participation in future profits. A basic proposal may offer to divide profits equally among all parties after the producer recoups the expenses of the production, which is an example of *net-profits participation*. Of course, what the expenses are and whether or not they will actually be recouped is a major question. It is rare in cable television that a profit participant is given a percentage of the gross receipts of the program.

There is nothing wrong with accepting a net-profits participation if the other elements of the compensation are acceptable. However, it is strongly urged that you not rely upon such provision as the major element of compensation. Regrettably, net profits frequently equal no profits. In the world of television, there is a long history of grievances on the part of countless net-profit participants who never received any profits—even for shows which have had successful network runs and then gone on to extended success in syndication (reruns by local stations). Remember: There will be no net profits until the original production costs are recouped, including your compensation, as well as the continuing costs of advertising, promotion, and distribution of the program.

Recoupment is all the more difficult because of the typical manner in which expenses are derived. For instance, it is not uncommon for the cable company to retain a certain percentage of any gross receipts received from exploitation of the program as a "distribution fee," which is the cable company's profit "off the top." Not uncommonly, the distribution fees received by a cable company will vary, depending upon the medium to which it is going to license the work. Typical fees are: 10 percent if the program is broadcast on network television in the United States; 20 percent for sale for television outside the United States; 15 percent for home video rights, that is, video disc and videocassette rights; and perhaps 17.5 percent for all other rights. These are average figures and are subject to negotiation and change, and they may change from company to company. Insure that there is a provision requiring fair and "arms-length" transactions to prevent a cable subsidiary of a network television company from licensing the program to its parent for less than fair value.

Distribution fees are calculated by the cable company before recoupment of the second element, distribution *expenses,* begins. Distribution expenses might include advertising, preparation of dubbed or titled versions of the work, additional laboratory costs for making new tapes of the program, freight, storage and insurance expenses, and possibly even the cost of any litigation which might have arisen out of the program. Finally, after it has recouped the distribution fees and distribution expenses, it will recoup the original cost of the production. At that point, theoretically, the profit participants would share in net profits. Given this formula, however, net profits are a long way down the road. If you are in a strong enough bargaining position, you would obviously want to reverse the above scheme, or perhaps try to put a cap on the cable company's distribution fees or the distribution expenses. Reversing the above scheme will result in much quicker recoupment of production costs, since all gross receipts are first applied to recoupment. If the distribution fee is subtracted from gross receipts first, there is a smaller amount to apply to recoupment. Remember, there are no net profits until recoupment.

Consider the example of a program which has a theoretical production cost of $100. Assume further that $300 of gross receipts are received from exploitation of the program, that the cable company is entitled to a 25 percent distribution fee, and has incurred $25 in distribution expenses, and that net profits are to be split 50/50 between the artist and the cable company. If the distribution fee is subtracted first, the cable company will receive $75 in profits off the top, which will leave $225 remaining. From that sum, the $25 in distribution expenses is subtracted, leaving $200. That $200 is then applied to the $100 production cost, so that the program will now have recouped. The remaining $100 will be split 50/50 between the cable company and the artist.

If the formula is reversed, the production costs are recouped first, from the same $300 of gross receipts, $100 will be subtracted, leaving $200. At this point, the program will have been recouped. From the $200 will be subtracted the $25 in distribution expenses leaving $175. Only at that time will the 25 percent distribution fee be applied, giving the cable company a profit of $37.50, which will be subtracted from the $175, leaving $137.50 to be split equally by the cable company and the artist. The artist will receive $68.75 as opposed to $50 under the above formula. It is also possible to negotiate for a guaranteed receipt of net profits, which might consist of minimum payments made over a certain period of time. It is even possible to negotiate a reversion of rights in the program if there is a failure to pay a certain percentage of net profits.

All of this depends upon your negotiating stance. Most

often, artists participating in a cable program have very little clout and, consequently, they are left to the accounting wizardry of the cable company. Given that the cable television field has so far generated more publicity than income, net profits cannot be counted on at this point in time. Many cable companies are losing money, and there is an incentive to spread the losses among all the programs on the schedule, so that no one involved with any one program is likely to benefit from his net-profits arrangement.

In the cable television field, net profits usually apply only to exploitation of the program outside of the cable telecast itself. That is, any net profits will stem from revenues received, for example, for license or sale of the program in foreign countries, or perhaps for its broadcast on network television. However, the cable company is likely to insist that the initial license fee it pays to you will give it the right to cablecast the show without your sharing in its profits from that use of the program.

There are other problems concerning net profits which should be noted. Your program may be bartered for a program owned by another company, in which case there will be no exchange of money. Stipulate that you must receive some compensation if this happens. Another possibility is that your program may be sold as part of a package made up of several programs. If your program is the most popular one in the package, the fee allocated to your show may not be fair. If you can, you should prohibit this possibility for your show in your contract with the cable company.

Frequently, the cable company will charge a *production fee*, sometimes referred to as an *executive supervisory fee*, which is an estimated cost of the company's overhead. The production fee may be either a stated percentage of the production budget — perhaps 10 percent — or a flat sum of money. Typically, this is just another way to delay paying net profits.

Sometimes it is possible, and may be desirable, to negotiate for the cable company to receive a production fee in lieu of interest, which the cable company may attempt to charge on the total sum it has spent for the production, at a rate usually above the prime rate. Because under such an arrangement this interest must be recouped before any net-profit percentages can be paid, note how important this is if production costs are the last element recouped, after distribution fees and distribution expenses. This means that the interest continues to run. The cable company will argue that it had to borrow this sum of money in order to finance the program, and that it is losing this rate of interest by investing in the show. This argument should be met by advising the company that it is in the business of producing cable television programs and that adding interest charges to production expenses is not acceptable.

Another abomination to be found in net-profits provisions is the accounting clause. Look closely to determine when you are to receive net-profits statements. Typically, the statements are rendered on a semi-annual basis as of June 30 and December 31, with payment to be made to net-profits participants ninety days thereafter. That means that revenue which is received on January 1 will be accounted for on June 30, and will be paid to the net-profits participants on September 31. That also means that the cable company will have use of your money for nine months' time. Although you may argue for interest during that period, the cable company will counter vociferously that its accounting department cannot render such statements, especially if the precedent of your case is followed by other profit participants. It is recommended to try to narrow the accounting period from semi-annual to quarterly, with payment to be made sixty days after the end of each quarter.

You may also wish to isolate certain types of net profits for accounting purposes. For instance, if the program is sold or licensed for a flat fee rather than, or in addition to, being subject to future royalty payments, you should receive your share of the flat fee within a certain limited period of time after its receipt by the cable company, notwithstanding the general accounting periods.

In conjunction with the accounting clause, there should be a clause giving you the right to audit the cable company's accounting books. This will require that you hire an accountant to unravel the complexity of the cable company's accounting system. Not infrequently, your right to audit will be limited to one or two years after receipt of any particular statement. Negotiate to expand this period. Frequently, several years will pass before a pattern of accounting abuse arises, causing you to question prior statements. But by that time, you right to audit many statements may have lapsed.

Finally, request a *most favored nation clause*. This requires the cable company to provide you with better terms for any of its standard provisions, in the event that subsequent to your agreement such provisions are changed to be more favorable to future participants in the cable company's programs. This should apply to all standard provisions of the agreement, although it is particularly important for the net-profits provisions because they contain so much standard verbiage. Rest assured that the cable company cannot use this clause to derogate any of the terms of your agreement. Nor does the clause apply to non-standard terms of your agreement, such as the amount paid to you as a licensing fee.

In summary, given the vagaries of net profits, it seems wiser to negotiate to receive a higher immediate compensation, rather than rely upon profit participations.

This chapter is meant as a basic introduction for a person who has never negotiated a cable deal before. It by no means exhausts all of the aspects of the negotiation process and should not be relied upon as a substitute for proper representation by persons knowledgeable about the cable television industry. It is only a primer which will allow the artist to better evaluate the deal, the contract, and the representation he is receiving.

Trade and Service Organizations

Association of Independent Video and Filmmakers, Inc. (AIVF). 625 Broadway, New York, NY 10012, (212) 473-3400.

AIVF is a nonprofit trade association dedicated to the promotion of independent video and film and to the effective national representation of independent producers.

Cabletelevision Advertising Bureau (CAB). 767 Third Avenue, New York, NY 10017, (212) 751-7770.

CAB is an industry trade association which provides sales and resource materials and helps to build advertising capability. It offers a monthly newsletter and consultation service. Members are composed of system operators, program services, sales representatives, and associate members.

The Cable Television Information Center (CTIC). 1800 North Kent Street, Suite 1007, Arlington, VA 22209, (703) 528-6846.

CTIC is a nonprofit membership organization which aims to help local officials make informed decisions regarding cable television. The information service provides educational publications and a monthly newsletter service, *CTIC Cable Reports*.

Independent Cinema Artists and Producers (ICAP). 625 Broadway, New York, NY 10012, (212) 522-9183.

ICAP is a nonprofit media arts association, founded in 1975 by independent filmmakers, which advises and aids independent film and video artists who desire representation in the field of electronic delivery systems.

Minorities in Cable. 105 Madison Avenue, New York, NY 10016, (212) 683-5656 x187.

Minorities in Cable is a nonprofit organization which serves as a forum to bring the minority community and the cable industry together to encourage positive growth for both.

National Assembly of Media Arts Centers (NAMAC). c/o New York Foundation for the Arts, 5 Beekman Street, Room 600, New York, NY 10038, (212) 233-3900.

NAMAC is a regionally organized, national alliance of media arts organizations and individuals designed to provide mutual assistance and to inform the public about the impact and value of independent film, video, and audio.

National Association of Telecommunications Officers and Advisors (NATOA). National League of Cities (NLC), 1301 Pennsylvania Avenue, NW, Washington, DC 20004, (202) 626-3115.

NATOA is a technical assistance division of the NLC, a lobbying group for cities across the U.S. NATOA is a networking organization of city cable regulators. It has annual meetings, publishes a newsletter, and deals with the issues of regulating cable and developing municipal uses of cable systems.

National Cable Arts Council (NCAC). P.O. Box 30498, New Orleans, LA 70190, (504) 837-4443.

NCAC is a communications network providing access to research, information, and consulting services.

National Cable Television Association (NCTA). 1724 Massachusetts Avenue, NW, Washington, DC 20036, (202) 775-3550.

NCTA is the trade association for the industry and its chief lobbying organization.

National Federation of Local Cable Programmers (NFLCP). 906 Pennsylvania Avenue, SE, Washington, DC 20003, (202) 544-7272.

NFLCP is a national, nonprofit membership organization which seeks to encourage development of community uses of cable channels. It hosts national and regional workshops and seminars on access, programming, management, and development; publishes newsletters; and provides fee-for-service consulting.

Office of Communication, Community Telecommunications Services, United Church of Christ. 105 Madison Avenue, New York, NY 10016, (212) 683-5656.

The Office of Communication has been involved in communications education since 1951. As a representative of the public interest in mass communications media, it has developed projects to: encourage equal employment opportunity in cable; assist churches and communities in negotiating strong cable franchises; train church and neighborhood leaders in ways to take advantage of cable; and provide resources to help people use cable effectively.

Women in Cable. 2033 M Street, NW, Suite 703, Washington, DC 20036, (202) 296-7245.

Women in Cable is a professional society for people in the cable industry which provides information and networking. There are twenty-one chapters and 1900 members nationally.

Bibliography

The following bibliography is a selected list of books and periodicals which were used by the author to develop the manuscript for this book.

Books

ABC. The *1981 Annual Report*. New York: American Broadcasting Companies, Inc., 1982.

A.C. Nielsen Company. *Glossary of Cable and TV Terms*. New York: A.C. Nielsen Company, 1981.

Arlen, Michael J. *Living-Room War*. New York: The Viking Press, 1969.

Ambrust, Sharon, ed. *The Cable TV Financial Databook*. December 1982. Carmel, CA: Paul Kagan, 1982.

_____. *The Cable TV Program Databook*. May 1983. Carmel, CA: Paul Kagan, 1982.

Beck, Kirsten, ed. *Cable Television and the Performing Arts*. New York: New York University, School of the Arts, 1981.

Beser, Stanley and Johnson, Leland. *An Economic Analysis of Mandatory Leased Channel Access for Cable Television*. Rand Corporation Report #R2989-MF. Santa Monica, CA: Rand Corporation, 1982.

Brown, Les. *Keeping Your Eye on Television*. New York: The Pilgrim Press, 1979.

_____. *The New York Times Encyclopedia of Television*. New York: New York Times Books, 1977.

Compaine, Benjamin M., ed. *Who Owns the Media?*. New York: Harmony Books, 1979.

Cunningham, John E. *Cable Television*. Second edition. Indianapolis, IN: Howard W. Sams & Co., Inc., 1981.

Eliot, T.S. *The Waste Land and Other Poems*. New York and London: Harcourt Brace Jovanovich, 1962.

Halberstam, David. *The Powers That Be*. New York: Dell Publishing Co., 1979.

Hollowell, Mary Louise, ed. *The Cable/Broadband Communications Book*. Vol. 2. Washington, DC: Communications Press, Inc., 1980-81.

_____. *The Cable/Broadband Communications Book*. Vol. 3. Washington, DC: Communications Press, Inc. 1982-83.

Jesuale, Nancy J., Neustadt, Richard M., and Miller, Nicholas P., eds. *A Guide for Local Policy*. The CTIC Cable Books, vol. 2. Arlington, VA: The Cable Television Information Center, 1982.

Jesuale, Nancy J. and Smith, Ralph Lee, eds. *The Community*

Medium. The CTIC Cable Books, vol. 1. Arlington, VA: The Cable Television Information Center, 1982.

Johnson, Nicholas. *How to Talk Back to Your Television Set.* Boston and Toronto: Little Brown and Company, 1970.

Kaatz, Ronald B. *Cable: An Advertiser's Guide to the New Electronic Media.* Chicago: Crain Books, 1982.

Kirk, Brian. "Cable Access in Boston: The Fight Over Governance." *Access.* April 1983, p.1.

Loeffler, Carl E. and Tong, Darlene, eds. *Performance Anthology: Source Book for a Decade of California Performance Art.* San Francisco: Contemporary Arts Press, La Mamelle, Inc., 1980.

Mander, Jerry. *Four Arguments for the Elimination of Television.* New York: Morrow Quill Paperback, 1978.

Mankiewicz, Frank and Swerdlow, Joel. *Remote Control: Television and the Manipulation of American Life.* New York: Times Books, 1978.

Manoff, Robert Karl, ed. *Cable Production: What Every Arts Organization Needs to Know.* New York: Volunteer Lawyers for the Arts, 1982.

Minow, Newton N. "The Vast Wasteland." In *Documents of American Broadcasting,* edited by Frank J. Kahn. Englewood Cliffs, NJ: Prentice-Hall, Inc., 1978.

The National Cable Television Association, Inc. *A Cable Primer.* Washington, DC: NCTA Association Affairs Department, 1981.

_____. *Careers in Cable.* Washington, DC: NCTA Association Affairs Department.

Peck, Diana. *The Cable Television Franchising Primer.* Washington, DC: National Federation of Local Cable Programmers, Inc., 1980.

Price, Monroe and Wicklein, John. *Cable Television: A Guide for Citizen Action.* Philadelphia: Pilgrim Press, 1972.

Rice, Jean, ed. *Cable TV Renewals and Refranchising.* Washington, DC: Communications Press, Inc., 1983.

Shaffer, William Drew and Wheelwright, Richard, eds. *Creating Original Programming for Cable TV.* White Plains, NY: Knowledge Industry Publications, Inc., 1983.

Smith, Ralph L. *The Wired Nation.* New York: Harper & Row, 1982.

Stearns, Jennifer. *A Short Course in Cable.* Sixth edition. New York: Office of Communication, United Church of Christ, 1981.

Television Audience Assessment, Inc. *Thje Audience Rates Television.* Cambridge, MA: Television Audience Assessment, Inc., 1983.

Television Factbook, Services Volume. Annual. Washington, DC: Television Digest.

Tyler, Ralph. *Changing Channels: The Story of Cable TV In America.* Norwalk, CT: On Cable Publications, Inc., 1982.

Van Dalsen, Randy. *Access Center Handbook.* Colorado: United Cable Television Corporation, 1981.

Wicklein, John. *Electronic Nightmare: The New Communications and Freedom.* New York: The Viking Press, 1979.

Periodicals

Access. Washington, DC: Telecommunications Research Action Center. Bi-weekly. Subscription information: P.O. Box 12038, Washington, DC 20005.

ArtCom. San Francisco: Contemporary Arts Press, La Mamelle. Quarterly. Subscription information: P.O. Box 3123, Rincon Annex, San Francisco, CA 94119.

Broadcasting Magazine. Washington, DC: Broadcasting Publications Inc. Weekly. Subscription information: 1735 De Sales Street, NW, Washington, DC 20036.

Broadcasting Yearbook. Washington, DC: Broadcasting Publications Inc. Subscription information: 1735 De Sales Street, NW, Washington, DC 20036.

Cable Marketing. New York: Jobson Publishing Corporation. Monthly. Subscription information: 352 Park Avenue South, New York, NY 10010.

CableVision. Englewood, CO: Titsch Communications. Weekly. Subscription information: P.O. Box 5727TA, Denver, CO 80217.

Channels of Communication. New York: Media Commentary Council Inc. Bi-monthly. Subscription information: Channels, Subscription Service Department, P.O. Box 2001, Mahopac, NY 10541.

Cable Television Business. Denver: Cardiff Publishing Co. Semimonthly. Subscription information: Cardiff Publishing, Circulation Service Center, P.O. Box 6229, Duluth, MN 55806.

Community Television Review. Washington, DC: National Federation of Local Cable Programmers. Quarterly. Subscription information: 906 Pennsylvania Avenue, SE, Washington, DC 20003.

Multichannel News: The Newspaper for the Electronic Media. New York: Fairchild Publications. Weekly (except the last week of December). Subscription information: P.O. Box 1124, Dover, NJ 07801.

View: The Magazine of Cable TV Programming. New York: View Communications Corporation. Monthly. Subscription information: 150 East 58th Street, New York, NY 10155.

About the American Council for the Arts

The American Council for the Arts (ACA) addresses significant issues in the arts by promoting communications, management improvement, and problem-solving among those who shape and implement arts policy.

ACA currently accomplishes this by:

- fostering communication and cooperation among arts groups and leaders in the public and private sectors;
- promoting advocacy on behalf of *all* the arts;
- sponsoring research, analysis, studies;
- publishing books, manuals, *American Arts* magazine and *ACA Update* for leaders and managers in the arts;
- providing information and clearinghouse services;
- providing technical assistance to arts managers and administrators.

Special thanks to the following for their contributions in support of ACA's operations, programs, and special projects.

BENEFACTORS

Aetna Life & Casualty Foundation
American Telephone & Telegraph Company·Atlantic Richfield Company
Edward M. Block·Marshall S. Cogan·Exxon Corporation
Gannett Foundation·Knoll International
National Endowment for the Arts·New York State Council on the Arts
Warner Communications Inc.

SUSTAINERS

Barbaralee Diamonstein-Spielvogel·Louis Harris & Associates
Charles Pfister·Rev. & Mrs. Alfred Shands III

PATRONS

All Brand Importers·American Express Company·Batus Inc.
Bendix Corporation·Browne-Vintners·Mr. & Mrs. Nicola Bulgari
CBS Incorporated·Chesebrough-Pond's Inc.·Disney Foundation
General Electric Foundation·IBM Corporation·ITT·John Kilpatrick, Jr.
Mobil Foundation, Inc.·Philip Morris Incorporated
R.J. Reynolds Industries, Inc.·Sakowitz, Inc.
Shell Companies Foundation
Mrs. Gerald H. Westby·Xerox Corporation

DONORS

Allied Foundation·The Allstate Foundation·Bankers Trust Company
Bristol-Myers Fund·The Chevron Fund·Donald G. Conrad
Cooper Industries·Dart & Kraft, Inc.·Marion V. Day
Emerson Electric Company·Estee Lauder Companies
Federated Department Stores, Inc.·Alan Feld
Ford Motor Company Fund·Gulf Oil Corporation
Heublein Foundation, Inc.·InterNorth Foundation
Manufacturers Hanover Trust·Monsanto Fund
NL Industries Foundation, Inc.·Norton Company Foundation
J.C. Penney Company, Inc.·Procter & Gamble Company

About Volunteer Lawyers for the Arts

Volunteer Lawyers for the Arts (VLA) arranges for free legal representation and counseling, and provides legal education to the arts community. Artists and nonprofit arts organizations with arts-related legal problems who are unable to afford private counsel are eligible for VLA's legal services.

VLA's publications and communications programs attempt to educate artists and their attorneys about the kinds of legal problems artists and arts organizations face, and to familiarize them with available solutions. VLA is a nonprofit, tax-exempt corporation supported by the National Endowment for the Arts, the New York State Council on the Arts, individual contributions, private foundations, and corporations.

Volunteer Lawyers for the Arts, 1560 Broadway, Suite 711, New York, NY 10036, (212) 575-1150. Arlene Shuler, Executive Director.